de Gruyter Textbook

Helmberg · Getting Acquainted with Fractals

Gilbert Helmberg

Getting Acquainted with Fractals

Walter de Gruyter
Berlin · New York

Gilbert Helmberg
Kalkofenweg 5
6020 Innsbruck
Austria

⊗ Printed on acid-free paper which falls within the guidelines
of the ANSI to ensure permanence and durability.

Library of Congress Cataloging-in-Publication Data

Helmberg, Gilbert.
Getting acquainted with fractals / by Gilbert Helmberg.
p. cm.
Includes bibliographical references.
ISBN 978-3-11-019092-2 (hardcover : alk. paper)
1. Fractals. I. Title.
QA614.86.H45 2007
514'.742–dc22
2006102211

Bibliographic information published by the Deutsche Nationalbibliothek

The Deutsche Nationalbibliothek lists this publication in the Deutsche Nationalbibliografie; detailed bibliographic data are available in the Internet at http://dnb.d-nb.de.

ISBN 978-3-11-019092-2

Printed in Germany.
Coverdesign: +malsy, kommunikation und gestaltung, Willich.
Printing and binding: Hubert & Co. GmbH & Co. KG, Göttingen.

Preface

To someone, having heard about fractals but not yet acquainted with them, they might seem to be regarded with suspicion: How could "real" objects – accessible by sight and not only by thought – be replicas of arbitrarily small parts of themselves? How could a continuous path which runs almost everywhere parallel to sea level climb up to any height? How could a continuous curve pass through every point of a square?

Getting acquainted with fractals opens a glimpse into a world of wonders, but these wonders are strongly supported by a frame of serious mathematics in which various of its branches play together: geometry, analysis, linear algebra, topology, measure theory, functions of complex variables, algebra, ...

I have tried to do justice to both aspects: the fascination of geometric objects as well as the serious mathematical background – as far as an advanced undergraduate level. At some points, where the technicalities would transgress this level, I have at least indicated where an interested reader could find the whole story. I hope the presentation adds something worthwhile to the many remarkable books on this topic which also lead much farther into the world of fractals.

These books also contain something which a reader might miss in the present one: I have chosen to avoid the possibility of frustrating the reader by expecting him to do exercises; he will find them in abundance in the mentioned books (e.g. [Barnsley, 1988], [Falconer, 1990]) if he wants to. However, it is at least my intention to make accessible – via the internet address `http://techmath.uibk.ac.at/helmberg` – the programs producing the illustrations, thus enabling the reader to create and play with fractals according to his own taste.

My thanks are due to the de Gruyter Publishing Company, in particular to Dr. Plato, for their interest in and support of this book project. My first book has been dedicated to my parents, my wife, and my two eldest children, but there are more people who mean very much to me. Therefore this book is dedicated

to Chri, Moni, and Mui.

Innsbruck, Cavalese, August 2006 Gilbert Helmberg

Contents

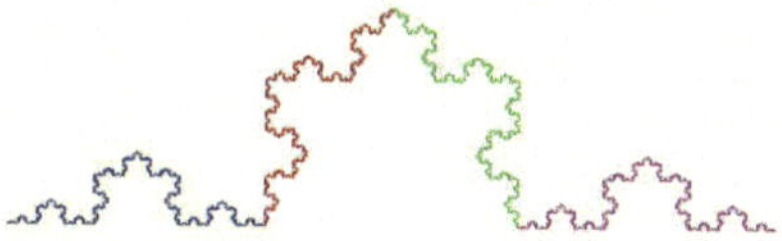

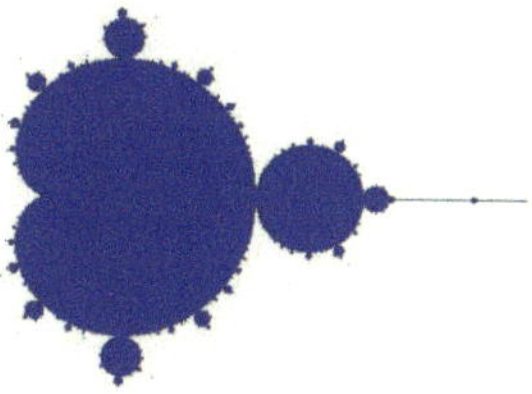

1 Fractals and dimension

1.1 The game of deleting and replacing

The word "fractal" comes from the Latin word "frangere" (with past participle "fractus") which means "to break", "to destroy". Let us begin with exploring how such a destruction process may still generate some new mathematical object displaying interesting features.

1.1.1 The CANTOR set

Let us define an operation f (such an operation is commonly called an *operator*) working on any closed segment $[a, b] \subset \mathbb{R}$ (= the real line) by deleting the open middle third $]a + \frac{b-a}{3}, b - \frac{b-a}{3}[$, and let us denote the interval $[0, 1] \subset \mathbb{R}$ by $A_{(0)}$. Application of f to $A_{(0)}$ deletes the interval $]\frac{1}{3}, \frac{2}{3}[$ and produces a closed set

$$A_{(1)} = \left[0, \tfrac{1}{3}\right] \cup \left[\tfrac{2}{3}, 1\right],$$

the union of the two disjoint closed intervals $A_0 = [0, \frac{1}{3}]$ and $A_1 = [\frac{2}{3}, 1]$, each of which has length $\frac{1}{3}$. If we apply f now to $A_{(1)}$ we get a closed set

$$A_{(2)} = f(A_{(1)}) = f(f(A_{(0)})) \subset A_{(1)}$$

consisting of four disjoint intervals $A_{0,0}$, $A_{0,1}$, $A_{1,0}$, $A_{1,1}$ of length $\frac{1}{9} = \frac{1}{3^2}$ each. Since we want to continue the application of f, in order to avoid the clumsy notation $f(f(\ldots))$ let us use the notation

$$\begin{aligned} f^{(0)}(A) &:= A, \\ f^{(1)}(A) &:= f(A), \\ f^{(k+1)}(A) &:= f(f^{(k)}(A)). \end{aligned}$$

(We shall call the index k the *level* of the construction.) Applied to our interval $A_{(0)}$ this allows us to define a sequence of closed sets $A_{(k)}$ $(1 \le k < \infty)$ by

$$A_{(k)} := f^{(k)}(A_{(0)})$$

satisfying

$$A_{(0)} \supset A_{(1)} \supset \cdots \supset A_{(k)} \supset A_{(k+1)} \supset \cdots . \tag{1.1}$$

The set $A_{(k)}$ is the union of 2^k closed intervals $A_{j_1,\ldots,j_k}$ $(j_i \in \{0, 1\},\ 1 \le i \le k)$ of length $\frac{1}{3^k}$ each. A sequence $\{A_{(k)}\}_{k=1}^{\infty}$ as well behaved as indicated by (1.1) raises the question whether there exists, in some sense, a limit set A. Indeed, by a well known topological theorem, the decreasing sequence of non-empty compact sets $\{A_{(k)}\}_{k=1}^{\infty}$ has the property that the set

$$A := \bigcap_{k=1}^{\infty} A_{(k)},$$

called the CANTOR *set* [Cantor, 1883], is compact and not empty. See Figure 1.1 for an illustration of the set $A_{(4)}$.

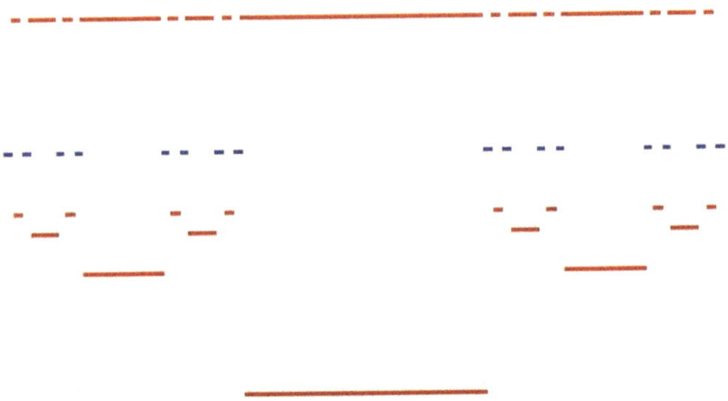

Figure 1.1. The set $A_{(4)}$, pictured in blue, is the union of sixteen closed component-intervals. The open set $[0,1] \setminus A_{(4)}$, pictured in red, is decomposed according to the intervals deleted at levels 1, 2, 3 and 4.

Still, as to the "size" of the set A, we notice that it is contained in all sets $A_{(k)}$ $(1 \le k < \infty)$; as observed above, the total length of the 2^k component-intervals of $A_{(k)}$ is $\frac{2^k}{3^k} = (\frac{2}{3})^k$ which approaches zero as $k \to \infty$. If A is to have any "length" in some sense at all, it therefore must be zero. Indeed, using one-dimensional LEBESGUE measure $\mathcal{L}$ $(= \mathcal{L}^1)$, which on the sets $A_{(k)}$ coincides with their lengths, by a well-known theorem of measure theory we get

$$\mathcal{L}(A) = \mathcal{L}\Big(\bigcap_{k=1}^{\infty} A_{(k)}\Big) = \lim_{k\to\infty} \mathcal{L}(A_{(k)}) = \lim_{k\to\infty} \Big(\frac{2}{3}\Big)^k = 0.$$

It is not surprising that A is not empty: the countably many end points of all component intervals $A_{j_1,\dots,j_k}$ $(0 \le k < \infty,\ j_i \in \{0,1\},\ 1 \le i \le k)$ are never deleted by any application of f and therefore are all contained in A. But there are more points surviving all these applications:

Let us write every non-zero $x \in [0,1]$ as an infinite series $x = \sum_{k=1}^{\infty} \frac{x_k}{3^k}$ $(x_k \in \{0,1,2\})$, in short $x = 0.x_1x_2\dots$ with the understanding that any finite sum of the form $x = 0.x_1x_2\dots 1 = \sum_{k=1}^{n} \frac{x_k}{3^k}$ $(x_n = 1)$ shall be written as a non-ending periodic triadic fraction

$$x = 0.x_1x_2\dots x_{n-1}022\dots = \sum_{k=1}^{n-1} \frac{x_k}{3^k} + \sum_{k=n+1}^{\infty} \frac{2}{3^k}.$$

Application of f to $A_{(0)}$ eliminates all points x for which $x_1 = 1$. The set $A_{(1)}$ therefore contains none of these. Renewed application of f to $A_{(1)}$ now eliminates all points x for which $x_2 = 1$ (in both intervals A_0 and A_1 which are characterized by $x_1 = 0$ and $x_1 = 2$ respectively). Repeated application of f subsequently eliminates all points $x \in A_{(0)}$ for which $x_k = 1$ $(1 \le k < \infty)$. What remains? Precisely the set of all points $x \in A_{(0)}$ whose "digits" x_k are either 0 or 2. It is well known that the points of the interval $[0, 1]$, apart from the countably many "dyadic rational" points, are in one-to-one correspondence with the points which in dyadic notation may be written as $y = 0.y_1y_2\dots$ $(y_k \in \{0, 1\})$. The conclusion is that our set A is not countable but contains as many points as the interval $[0, 1]$, i.e. has the cardinality of the reals.

A mathematician may be tempted to exploit the relation between the set A and a subset of $[0, 1]$ even further. Just now we have associated with the point

$$y = \sum_{k=1}^{\infty} \frac{y_k}{2^k} \qquad \Big(y_k \in \{0, 1\},\ \sum_{y_k=0} 1 = \sum_{y_k=1} 1 = \infty\Big)$$

the point

$$a(y) = \sum_{k=1}^{\infty} \frac{2y_k}{3^k} \in A.$$

Denoting by $\mathbb{N}$ the set of natural numbers, we may extend this mapping a to the compact topological product $\{0, 1\}^{\mathbb{N}}$ of all $\{0, 1\}$-sequences by defining

$$\tilde{a}(\tilde{y}) := \sum_{k=1}^{\infty} \frac{2y_k}{3^k} \qquad \text{for } \tilde{y} = \{y_k\}_{k=1}^{\infty} \quad (y_k \in \{0, 1\}),$$

e.g. if $\tilde{y} = \{0, 1, 1, \dots\}$, then $\tilde{a}(\tilde{y}) = \sum_{k=1}^{\infty} \frac{2}{3^k} = \frac{1}{3}$, while for $\tilde{y} = \{1, 0, 0, \dots\}$ we get $\tilde{a}(\tilde{y}) = \frac{2}{3}$. It is not hard to see that the mapping $\tilde{a} : \{0, 1\}^{\mathbb{N}} \to A$ is bijective (every point of A is the image of exactly one sequence in $\{0, 1\}^{\mathbb{N}}$) and continuous. A well-known topological theorem (cf. [Kelley, 1955, p. 141]) then asserts that A is homeomorphic to $\{0, 1\}^{\mathbb{N}}$. In particular, A is completely disconnected (in topology such a space is also called *zero-dimensional*) and *perfect* (i.e. closed without isolated points), but nowhere dense. Remembering that $[0, 1]$ may be considered as a subset of $\{0, 1\}^{\mathbb{N}}$ and roughly speaking, the mapping $\tilde{a}$ furnishes an extended parametrization of the set A (i.e. to every dyadic rational point of $[0, 1]$ there correspond two "neighbouring" points of A).

At this point we may notice one more property of the set A which is important for us since it will turn up in adapted form repeatedly in sets which we legitimately may call "fractals": suppose we omit the component A_1 of the set $A_{(1)}$ and restrict repeated application of f to the interval A_0. What would we have got? Evidently part of A, to wit a copy of the set A, only reduced by a factor $\frac{1}{3}$ in size. In fact, every component set $A_{j_1,\dots,j_k}$ of $A_{(k)}$, treated by itself with successive applications of f, produces a set which is part of A and, at the same time, a copy similar to A but reduced by a factor $\frac{1}{3^k}$. In other words, one may say that the set A is "self-similar" in the sense that it consists of smaller parts which are still similar to A.

A slight adaption of our construction of the CANTOR set furnishes a set with strikingly different features. Since we are now going to move from $\mathbb{R}$ to $\mathbb{R}^2$, and more generally to $\mathbb{R}^n$, for any vector $x = (x_1, \ldots, x_n) \in \mathbb{R}^n$ we shall use the EUCLIDEAN norm $|x| = \sqrt{\sum_{k=1}^{n} x_k^2}$. The (EUCLIDEAN) distance of two points $a \in \mathbb{R}^n$ and $b \in \mathbb{R}^n$ is then given by $|a - b|$.

1.1.2 The KOCH curve

We shall now modify the operator f considered in Section 1.1.1 by allowing it to work on any closed segment $[a, b]$ in the plane $\mathbb{R}^2$, and in the following way: it not only deletes but replaces the open middle third $]a + \frac{b-a}{3}, b - \frac{b-a}{3}[$ by two sides of an equilateral triangle, side length $|\frac{b-a}{3}|$, located to the left of $[a, b]$ if this segment is directed from a to b. We shall apply f also to piecewise linear curves in $\mathbb{R}^2$. Such a curve E is the graph of a piecewise linear, not necessarily continuous, function $g : [0, 1] \to \mathbb{R}^2$. It consists of finitely many segments $[a_j, b_j]$ $(1 \leq j \leq n)$, at most pairwise joined at their endpoints. The result $f(E)$ of applying f to E is obtained by applying the operator f to each of the component segments of E.

We start out again with the segment $A_{(0)} = [0, 1]$ on the x-axis. Application of f to $A_{(0)}$ produces a continuous piecewise linear curve $A_{(1)} = f(A_{(0)})$ consisting of four segments denoted consecutively by A_j $(0 \leq j \leq 3)$, each of which has length $\frac{1}{3}$ (Figure 1.2).

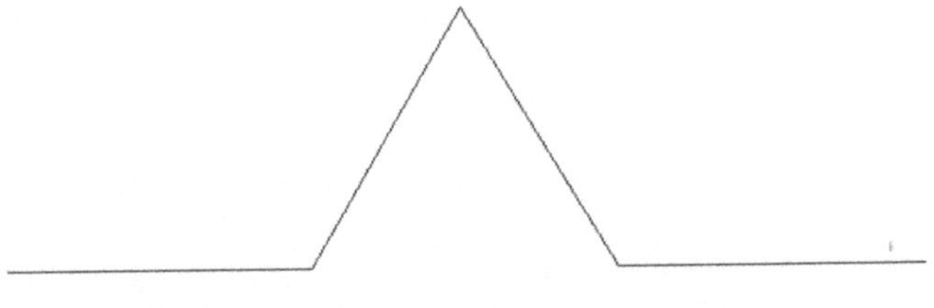

Figure 1.2. The generator $A_{(1)}$ of the KOCH curve.

Why not apply f again, this time to $A_{(1)}$, i.e. to each of these four segments? The result is a continuous piecewise linear curve $A_{(2)} = f^{(2)}(A_{(0)})$ consisting of 4^2 segments A_{j_1,j_2} $(0 \leq j_i \leq 3)$ of length $\frac{1}{3^2}$ each. Repetition of this procedure furnishes a sequence of continuous piecewise curves $A_{(k)}$ consisting of 4^k segments $A_{j_1,\ldots,j_k}$ $(0 \leq j_i \leq 3,\ 1 \leq i \leq k)$ of length $\frac{1}{3^k}$ each (Figures 1.3–1.5). Unfortunately, however, these curves, considered as subsets of $\mathbb{R}^2$, do not anymore satisfy (1.1). Do they still converge in some sense to some limit? The eye emphatically approves, but does mathematics support this impression?

In order to investigate the situation, we turn our attention to the sequence of segments $A_{j_1,\ldots,j_k}$ in $\mathbb{R}^2$ constituting the curve $A_{(k)}$ successively, starting at $(0, 0)$ and ending at $(1, 0)$. Notice that the endpoints of these segments are preserved when f

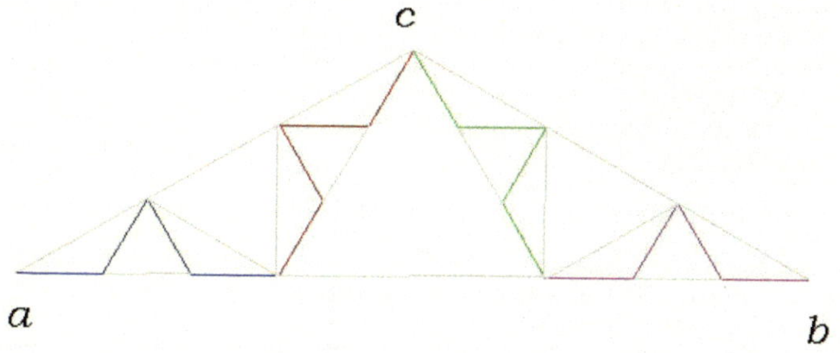

Figure 1.3. The approximating set $A_{(2)}$ for the KOCH curve. The grey lines illustrate the open set condition (Definition 1.1.3.2) needed for the computation of $\dim_S(A)$.

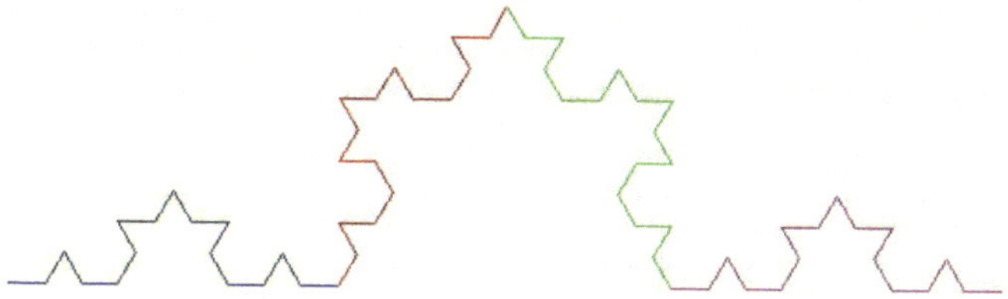

Figure 1.4. The approximating set $A_{(3)}$ for the KOCH curve.

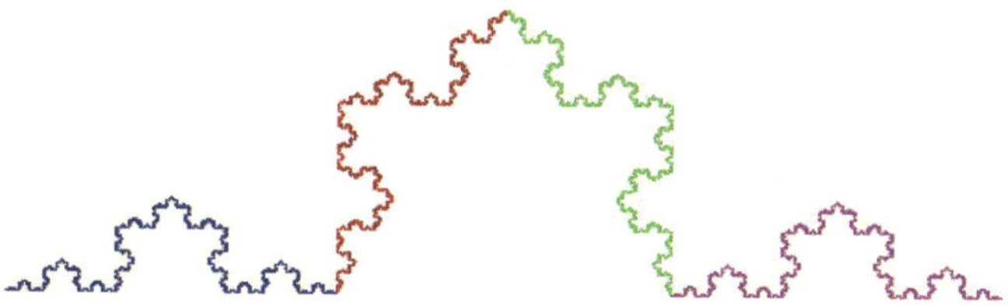

Figure 1.5. A closer approximation ($A_{(7)}$) of the KOCH curve A. The four colours indicate subsets of A which are similar to the whole of A.

is applied to $A_{(k)}$ since they only become endpoints of smaller subsegments. Let us define a map $\phi_k : [0,1] \to A_{(k)}$ in the following way: write every $x \in [0,1]$ in its "4-adic expansion"

$$x = \sum_{i=1}^{\infty} \frac{x_i}{4^i} = \sum_{i=1}^{k} \frac{x_i}{4^i} + r_k(x) \qquad (x_i \in \{0,1,2,3\}, 1 \le i < \infty). \tag{1.2}$$

If we agree to use the finite sum expansion when possible, then $0 \le r_k(x) < \frac{1}{4^k}$. Now define $\phi_k(x)$ to be the point of the segment $A_{x_1,x_2,\dots,x_k}$ lying at distance $r_k(x)$ from the starting point of this segment (in the positive direction). Evidently ϕ_k is a continuous piecewise linear map of $[0,1]$ onto $A_{(k)}$.

What happens to ϕ_k if k increases to ∞? In order to find out about this let us observe what happens to the point $\phi_k(x)$ if x is given as in (1.2) and if we define

$$x_{(m)} := \sum_{i=1}^{m} \frac{x_i}{4^i}.$$

As pointed out above, for the "4-adic rational" part $x_{(m)}$ of x and for all $k \ge m$ we get

$$\phi_k(x_{(m)}) = \phi_m(x_{(m)}) \in A_{x_1,\dots,x_m,0,\dots,0;} \subset A_{(k)} \tag{1.3}$$

(in fact, $\phi_k(x_{(m)})$ is the starting point of this subsegment of $A_{(k)}$). Observing the effect of consecutive applications of f to $A_{x_1,\dots,x_m}$, we find that every such application moves the point $\phi_k(x) \in A_{x_1,\dots,x_m,\dots,x_k}$ to its new position $\phi_{k+1}(x) \in A_{x_1,\dots,x_m,\dots,x_k,x_{k+1}}$ about a distance of at most $\frac{4}{3^{k+1}}$ (four times the length of $A_{x_1,\dots,x_{k+1}}$; a rough estimate since at most $\frac{\sqrt{3}}{2 \cdot 3^{k+1}}$ would do). Adding this up for $k > m$ we get the estimate

$$|\phi_k(x) - \phi_m(x)| \le \sum_{i=m}^{k-1} \frac{4}{3^{i+1}} < \sum_{i=m}^{\infty} \frac{4}{3^{i+1}} = \frac{2}{3^m}. \tag{1.4}$$

As a consequence, we see that the sequence $\{\phi_k(x)\}_{k=1}^{\infty}$ is a CAUCHY sequence (= fundamental sequence) in the plane $\mathbb{R}^2$ and has to converge to a limit point $\phi(x)$. By (1.4) we even see that the functions ϕ_k converge uniformly on $[0,1]$ and that therefore the limiting map ϕ furnishes a continuous curve in $\mathbb{R}^2$. This curve is called the KOCH *curve* [von Koch, 1904].

The fact that endpoints of subsegments $A_{x_1,\dots,x_m}$ of $A_{(m)}$ do not change position under further applications of f, as expressed by (1.3), helps to realize that the KOCH curve is nowhere differentiable. Non-differentiability is readily seen at such an endpoint itself: For any $x \in [0,1[$ consider the point $p_0 = \phi(x_{(m)})$ as defined above (for $x = 1$ the reasoning has to be slightly adapted). For $k > m$ let

$$p_1 := \phi\Big(x_{(m)} + \frac{1}{4^k}\Big), \qquad p_2 := \phi\Big(x_{(m)} + \frac{2}{4^k}\Big).$$

p_1 is the endpoint of the subsegment of $A_{(k)}$ beginning at $\phi(x_m)$, p_2 is the endpoint of the following subsegment of $A_{(k)}$. If $k > m$ is sufficiently large, then the points p_1 and

p_2 are arbitrarily close to p_0, while the secants p_0p_1 and p_0p_2 always include the same positive angle.

It is somewhat more tedious to deal with a point $\phi(x)$ if x is not a 4-adic rational number. Roughly speaking, if ϕ were differentiable in x, then two different points close to $\phi(x)$ (as which will be taken endpoints of subsegments) would have to define a secant close to the tangent in $\phi(x)$, and this will be shown to be impossible. Recall that for complex-valued functions g and h of the argument y one writes $h = o(g)$ as $y \to a$ if $\lim_{y\to a} \frac{h(y)}{g(y)} = 0$, while $h = O(g)$ as $y \to a$ means $\limsup_{y\to a} \frac{|h(y)|}{|g(y)|} < \infty$. Correspondingly $o(1)$ (as $y \to x$) will denote a function which vanishes as $y \to x$, and $O(1)$ will denote a function which remains bounded as $y \to x$. Although not strictly necessary the notation $o_1(1), o_2(1), \ldots, O_1(1), O_2(1), \ldots$ will be used to indicate different such functions.

1.1.2.1 Lemma. *Suppose ϕ is differentiable in x, i.e. there exists a vector $q \in \mathbb{R}^2$ such that*

$$\phi(x) - \phi(y) = (q + o(1)) \cdot (x - y) \quad \text{as } y \to x,$$

and let $y_1 \to x$ and $y_2 \to x$ in such a way that

$$\begin{aligned} x - y_1 &= O_1(y_1 - y_2), \\ x - y_2 &= O_2(y_1 - y_2). \end{aligned} \tag{1.5}$$

Then

$$\phi(y_1) - \phi(y_2) = (q + o(1)) \cdot (y_1 - y_2). \tag{1.6}$$

Proof of the lemma.

$$\begin{aligned} \phi(y_1) - \phi(y_2) &= \phi(x) - \phi(y_2) - (\phi(x) - \phi(y_1)) \\ &= (q + o_2(1)) \cdot (x - y_2) - (q + o_1(1)) \cdot (x - y_1) \\ &= q \cdot (y_1 - y_2) + o_2(1) \cdot (x - y_2) + o_1(1) \cdot (x - y_1) \\ &= [q + o_2(1) \cdot O_2(1) + o_1(1) \cdot O_1(1)] \cdot (y_1 - y_2). \end{aligned}$$

□

In order to show that ϕ cannot be differentiable in x let, for arbitrarily large m,

$$y_1 := x_{(m)}, \qquad y_2 := x_{(m)} + \frac{1}{4^{m+1}}, \qquad \overline{y_2} := x_{(m)} + \frac{2}{4^{m+1}}.$$

Then

$$\begin{aligned} |y_1 - \overline{y_2}| &= 2 \cdot |y_1 - y_2| = \frac{2}{4^{m+1}}, \\ |x - y_1| &\le \frac{4}{4^{m+1}} = 4 \cdot |y_1 - y_2|, \\ |x - y_2| &\le \frac{3}{4^{m+1}} = 3 \cdot |y_1 - y_2|, \\ |x - \overline{y_2}| &\le \frac{2}{4^{m+1}} = |y_1 - \overline{y_2}|. \end{aligned}$$

So the requirements (1.5) are satisfied, but not (1.6) since we have already seen that the secants $\phi(y_1)\phi(y_2)$ and $\phi(y_1)\phi(\overline{y_2})$ always include the same small but non-zero angle. Consequently, the function ϕ cannot be differentiable in x.

How do we measure the length of a continuous curve? Take any finite sequence $S = \{p_j\}_{j=0}^{n}$ of points corresponding to increasing parameter values and compute $l_S := \sum_{j=1}^{n} |p_j - p_{j-1}|$. The length of the curve is then by definition the supremum over all values l_S obtained in this way. For the KOCH curve it seems convenient to choose $S_k := \{\phi(\frac{j}{4^k})\}_{j=0}^{4^k}$, the endpoints of the 4^k subsegments in $A_{(k)}$. Each of these has length $\frac{1}{3^k}$, therefore we get $l_{S_k} = (\frac{4}{3})^k$. As $k \to \infty$ this also tends to ∞. We conclude that the KOCH curve has infinite length, rather a contrast to the CANTOR set.

One last question (for the time being): what would have happened if we had restricted the action of f to one subsegment A_j ($j \in \{0, 1, 2, 3\}$) of $A_{(0)}$ or, more generally, to a subsegment $A_{j_1,\dots,j_k}$ of $A_{(k)}$? Obviously we would have got a curve similar to A but reduced to $\frac{1}{3}$, resp. $\frac{1}{3^k}$, in size. In other words, again the KOCH curve is self-similar, it consists of parts which are smaller copies of itself.

1.1.3 Heuristics of dimension

We have not yet pinned down any property of a set in $\mathbb{R}^2$ or $\mathbb{R}^3$ or, more generally, $\mathbb{R}^n$ which might be strange and characteristic enough to make it reasonable to call the set a "fractal". Self-similarity as encountered in the CANTOR set or the KOCH curve seems a possible candidate but there are perfectly harmless sets which also are self-similar, for instance a square in the plane. Shrinking its sides to half the original length again produces a square and the original square consists of four copies thereof – if we allow the sides of the small squares to coincide. In fact, this is intimately connected with the assertion that a full square is a set of dimension 2: reducing the sides to $\frac{1}{n}$ of their original length produces a set, n^2 copies of which (allowing sides to coincide) constitute the original square. Similarly n^3 cubes of side length $\frac{1}{n}$ make up the unit cube – corresponding to its three-dimensionality – and n^1 intervals of length $\frac{1}{n}$ joined together give the one-dimensional unit interval.

If we had not already been familiar with the concept of dimension we could have "computed" the dimension of a square A, say, using the following reasoning: the dimension of A is the exponent d determined by the fact that the set A is a ("almost disjoint", whatever this may mean) union of n^d similar copies of A, reduced in size by the factor $\frac{1}{n}$ (such a similar copy $S(A)$ is congruent with the set $\frac{1}{n}A$, which originates by multiplying every vector in A by the factor $\delta(S) = \frac{1}{n}$). In other words and roughly speaking (to be made more precise in later sections), if $\delta(S) = \frac{1}{n}$ and if A happens to be decomposable into $N_S(A)$ sets of the form $\delta(S)A$, then the *self-similarity dimension* $\dim_S(A)$ may be considered as the solution of the equation

$$\left(\frac{1}{\delta(S)}\right)^{\dim_S(A)} = N_S(A),$$

i.e.

$$\dim_S(A) = \frac{\log N_S(A)}{-\log \delta(S)}. \tag{1.7}$$

Applying this reasoning to the CANTOR set A we recall that it is indeed the disjoint union of two similar copies of itself, reduced by the factor $\delta(S) = \frac{1}{3}$. Formula (1.7) now gives for its dimension

$$\dim_S(A) = \frac{\log 2}{\log 3} \approx 0.63.$$

There is one objection to be dealt with: $\dim_S(A)$ has not been defined in a unique way. If A is the (almost) disjoint union of $N_S(A)$ copies of $\delta(S)A$, then $\delta(S)A$ is the (almost) disjoint union of $N_S(A)$ (almost) disjoint copies of $(\delta(S))^2 A$ and A is the (almost) disjoint union of $N_{S^{(2)}}(A) = (N_S(A))^2$ copies of $\delta(S^{(2)})A = (\delta(S))^2 A$. Should we have been told, before applying formula (1.7), whether to work with S or with $S^{(2)}$, or even with the k-fold iteration $S^{(k)}$ of S? Fortunately this does not matter, since

$$\frac{\log N_{S^{(k)}}(A)}{\delta(S^{(k)})} = \frac{k \log N_S(A)}{k \log \delta(S)} = \frac{\log N_S(A)}{\log \delta(S)}.$$

If there is some doubt left, please be patient until dimension is discussed more thoroughly in Section 1.2 and Section 1.3.

The startling fact is that the dimension of the CANTOR set, with this understanding, is not 1 but less, to wit approximately 0.63 (also different from its topological dimension as a completely disconnected set, which is zero). Looking now at the KOCH curve, formula (1.7) tells us that (if the points in which the subsegments join do not do any damage) its self-similarity dimension is $\frac{\log 4}{\log 3} \approx 1.26$, while of course the topological dimension of each of its approximating sets $A_{(k)}$ is 1.

A theorem (Theorem 1.3.8) to be stated later tells a condition under which this reasoning is applicable, the so-called *open set condition*. Let us first state explicitly what is meant by a similarity.

1.1.3.1 Definition. A map $S : \mathbb{R}^n \to \mathbb{R}^n$ is called a *similarity* with *similarity factor* s if $|S(x) - S(y)| = s \cdot |x - y|$ for some positive number s and for all $x \in \mathbb{R}^n$ and $y \in \mathbb{R}^n$.

1.1.3.2 Definition. The similarities S_i $(1 \le i \le k)$ satisfy the *open set condition* if there exists a bounded non-empty open set V with mutually disjoint image sets $S_i(V)$ $(1 \le i \le k)$ satisfying $\bigcup_{i=1}^k S_i(V) \subset V$.

In essence the mentioned theorem states that if the similarities S_i $(1 \le i \le k)$ satisfy the open set condition and if $A = \bigcup_{i=1}^k S_i(A)$, then (1.7) and even a more general formula for the computation of $\dim_S(A = \bigcup_{i=1}^k S_i(A))$ may be applied. The open set condition is obviously satisfied in the case of the CANTOR set: denoting by S_1 and S_2 the similarities mapping the unit interval into its first and last third, as the set V we may take e.g. the open unit interval. It is satisfied also in case of the KOCH curve: let V be the open isosceles triangle with vertices $a = (0,0)$, $b = (1,0)$, $c = (\frac{1}{2}, \frac{1}{2\sqrt{3}})$ (see Figure 1.3). V contains its four images under the similarities mapping the unit interval into the four line segments constituting the set $A_{(1)}$. As a consequence, we may also note that the whole set A is contained in the closure of the triangle V.

Are we now in position to define what is meant by a "fractal"? Yes and no. Yes, since the original definition of MANDELBROT [Mandelbrot, 1982, Section 3] says: A subset of $\mathbb{R}^n$ is a *fractal* (also *fractal set*) if its topological dimension (which is always an integer, zero for the CANTOR set and one for the KOCH curve) is less than its "fractal" dimension (for the CANTOR set and the KOCH curve as computed above). According to this definition a set with a non-integral dimension (as discussed more generally later) is automatically a fractal. No, since it has turned out that there are sets (as the dragon Section 1.1.5.3 to be discussed later) that one would like to consider as fractals but are not included by the just mentioned definition. Up to now it has seemed difficult to find a satisfying definition including all sets which one would like to consider as fractals.

1.1.4 Initiators and generators

There are evidently two ways to produce more general fractals besides the CANTOR set and the KOCH curve: we can start from a set different from the unit interval $A = [0, 1]$ and we can (as we have done already) change the definition of the map f. The first is done by defining a set of segments $A_{(0)}$, then called the *initiator*, upon which the iterates $f^{(k)}$ of f should act. An example is provided as follows.

1.1.4.1 The KOCH island

Let $A_{(0)}$ be the equilateral triangle below the x-axis, one side of which is the unit interval $[0, 1]$. Applying the map f defining the KOCH curve (Section 1.1.2) to the three segments constituting the set $A_{(0)}$ produces a star with six vertices which we may also imagine as an equilateral hexagon, each side of which carries an equilateral triangle of side length $\frac{1}{3}$ (Figure 1.5). The next application of f adds twelve smaller equilateral triangles of side length $\frac{1}{9}$. Continuing this procedure eventually produces the contour of a set looking like a snow flake, consisting of three copies of the KOCH curve we know from Section 1.1.2. The idea of it being surrounded by water leads to calling it the KOCH *island* (Figures 1.6, 1.7).

Figure 1.6. The first approximating set $A_{(1)}$ for the KOCH island.

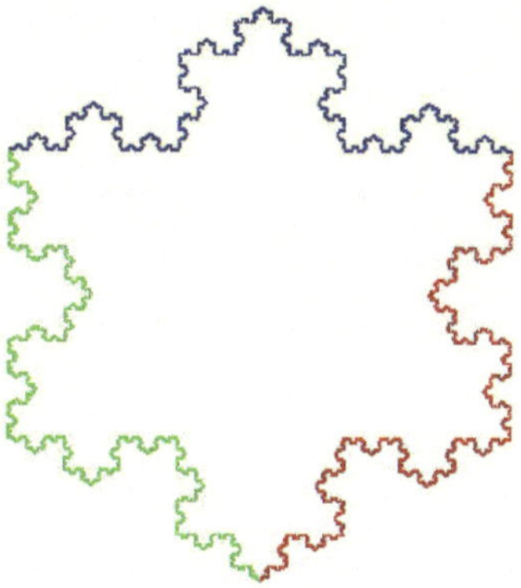

Figure 1.7. A closer approximation ($A_{(7)}$) of the KOCH island.

Still, there is more to this: somebody seeing it for the first time and being asked to estimate the length of the coast line may think: “Well, a little bit more than the perimeter of a circle roughly the same size; taking into account the coasts of the peninsulas and the bays, perhaps twice this perimeter.” Asked to look a bit closer and perhaps to use a compass with a rather small opening he may to his surprise find that his measurement of the coast line becomes longer and longer as he decreases this opening, until we disclose to him that already one third of the coast line – our well-known KOCH curve – has infinite length.

It is in this line of thought that MANDELBROT [Mandelbrot, 1982, Section 5] points out that also e.g. the coast line of England, measured with increasing precision, turns out to have infinite length.

Keeping, for the time being, the unit interval as our initiator $A_{(0)}$, we may change the mapping f by requiring that it should act on every segment of any union B of segments by replacing this segment with a – suitably diminished – similar copy of a given union G of segments, called the *generator*. Let us look at several samples of the vast family of fractals obtained in this way.

1.1.4.2 A modified KOCH curve

Suppose the generator G consists of five segments of length $\frac{1}{3}$ each, obtained by replacing the two middle segments of the KOCH curve generator with three sides of a square (Figure 1.8).

We may think of f as employing five similarity maps S_i $(1 \leq i \leq 5)$ each with similarity factor $\frac{1}{3}$. Now the fractal $A = \lim_{k\to\infty} f^{(k)}(A_{(0)})$, defined in essentially the same way as in Section 1.1.2, consists of the union of five similar copies S_iA $(1 \leq i \leq 5)$ joined at the vertices of the original generator $G = A_{(1)}$, but the first two and the last two copies having a lot more points (in fact a whole diagonal segment) in common (Figure 1.9). Still, the open set condition (Definition 1.1.3.2) is satisfied: the open isosceles right-angled triangle D with $A_{(0)}$ as hypotenuse contains the union

Figure 1.8. The generator of the first modified KOCH curve. The grey lines delimit the triangle D providing for the open set condition.

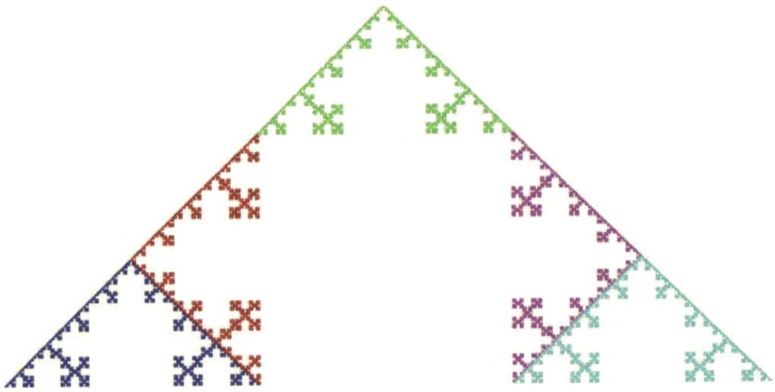

Figure 1.9. A closer approximation ($A_{(6)}$) of the first modified KOCH curve.

$\bigcup_{i=1}^{5} S_i D$ of its five copies under the similarities S_i $(1 \leq i \leq 5)$. According to what has been said in Section 1.1.3, we may compute the dimension of the set A as being $\dim_S(A) = \frac{\log 5}{\log 3} \approx 1.46$.

Suppose we still adapt the generator G by shortening its second and fourth segment to $\frac{1}{5}$, say, in place of $\frac{1}{3}$ (Figure 1.10).

The resulting fractal A (Figure 1.11) still consists of five similar copies of A, but the first, third, and fifth similarity employ the similarity factor $\frac{1}{3}$, while the second and fourth copy employ the similarity factor $\frac{1}{5}$. Now the first and second copies, as well as the fourth and fifth copies, have only the connecting vertices of the original generator in common, and the open set condition still holds since the former right-angled triangle above may be replaced by an oblique-angled symmetric triangle through the "outer" vertices of the new generator. Our method of determining a dimension of the resulting fractal A, however, breaks down, since the similar copies of A constituting A do not have the same size. If we consider A to be eligible for a dimension – as seems to be sensible – then we shall have to use a refined reasoning. Fortunately this will be

Figure 1.10. The generator of the second modified KOCH curve.

Figure 1.11. A closer approximation ($A_{(6)}$) of the second modified KOCH curve.

supplied by Theorem 1.3.8. An application thereof, using the concept of "HAUSDORFF dimension" will furnish the result $\dim_H(A) \approx 1.27$.

A more troublesome situation happens if we adapt the generator G by enlarging (instead of shortening) its second and fourth segment to $\frac{1}{2}$, say (Figures 1.12–1.14).

The second and fourth copy of the resulting fractal A appearing as part of A then employ the similarity factor $\frac{1}{2}$ and become entangled with (i.e. intersect) the first and fifth copy respectively to an extent which is not anymore negligible and no longer allows one to find an open set furnishing the open set condition. But should this prevent the fractal A from legitimately having a dimension? This would be very unsatisfactory

Figure 1.12. The generator of the third modified KOCH curve.

Figure 1.13. The approximating set $A_{(4)}$ for the third modified KOCH curve.

Figure 1.14. A closer approximation ($A_{(7)}$) of the third modified KOCH curve.

indeed. In Section 1.2 and Section 1.3 we shall therefore discuss two reasonable concepts of "dimension" which, although not always computable in practice, will give every reasonable set a number as its dimension.

1.1.4.3 A second type of modified KOCH curves

Let us consider a whole family of related fractals, starting with the unit interval as initiator and the generator G_1 consisting of eight segments of equal length joining consecutively the points

$$\begin{aligned} p_0 &= (0,0) & p_1 &= (0.25,0) & p_2 &= (0.25,0.25) \\ p_3 &= (0.5,0.25) & p_4 &= (0.5,0) & p_5 &= (0.5,-0.25) \\ p_6 &= (0.75,-0.25) & p_7 &= (0.75,0) & p_8 &= (1,0). \end{aligned}$$

The open parallelogram formed by the straight lines p_0p_2, p_3p_8, p_8p_6, p_5p_0 serves to verify the open set condition; the dimension of the resulting fractal is $\dim_S(A) = \frac{\log 8}{\log 4} = \frac{3}{2}$ (Figures 1.15, 1.16).

Replacing in the generator G_1 the two squares above and below the x-axis by equilateral triangles produces a generator G_2 consisting of six segments of equal length joining consecutively the points

$$\begin{aligned} p_0 &= (0,0) & p_1 &= (0.25,0) & p_2 &= (0.375, \tfrac{\sqrt{3}}{8} \approx 0.2165) \\ p_3 &= (0.5,0) & p_4 &= (0.625, -\tfrac{\sqrt{3}}{8}) & p_5 &= (0.75,0) \\ p_6 &= (1,0). \end{aligned}$$

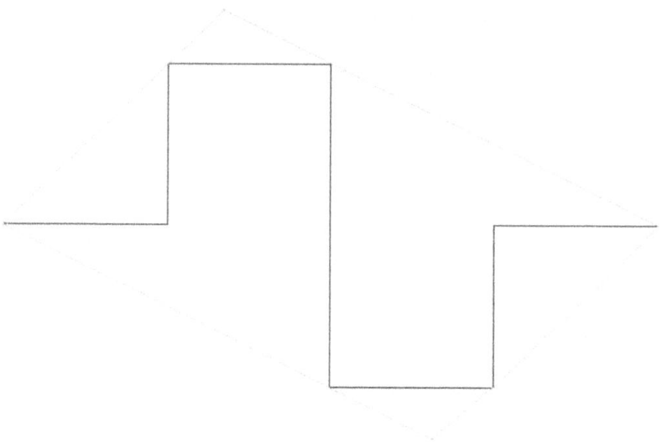

Figure 1.15. The generator G_1. The grey lines delimit the parallelogram providing for the open set condition.

Figure 1.16. The approximating set $A_{(5)}$ for the fractal with generator G_1.

Again the open parallelogram with vertices p_0, p_2, p_6, p_4 serves to verify the open set condition (also its similar images with diagonals p_3p_4 and p_4p_5 have only the point p_4 in common); the dimension of the resulting fractal is $\dim_S(A) = \frac{\log 6}{\log 4} \approx 1.29$ (Figures 1.17, 1.18).

It would be perfectly legitimate to omit the vertex p_3 and to consider p_2p_4 as a single segment, but formula (1.7) could no longer be used for the computation of the dimension of the resulting fractal.

Finally we contract the two truncated equilateral triangles in G_2 above and below the x-axis to segments of length $\frac{1}{3}$, positioned perpendicularly to the unit interval at its points $(\frac{1}{3}, 0)$ (above the x-axis) and $(\frac{2}{3}, 0)$ (below the x-axis) respectively (G_3). In fact although each approximating curve $f^{(k)}(A_{(0)})$ runs twice through these segments, because of the symmetry properties of the generator we may now envision the resulting

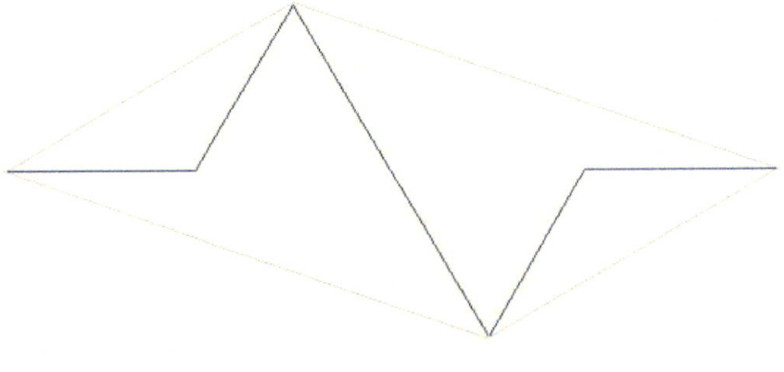

Figure 1.17. The generator G_2. The grey lines delimit the parallelogram providing for the open set condition.

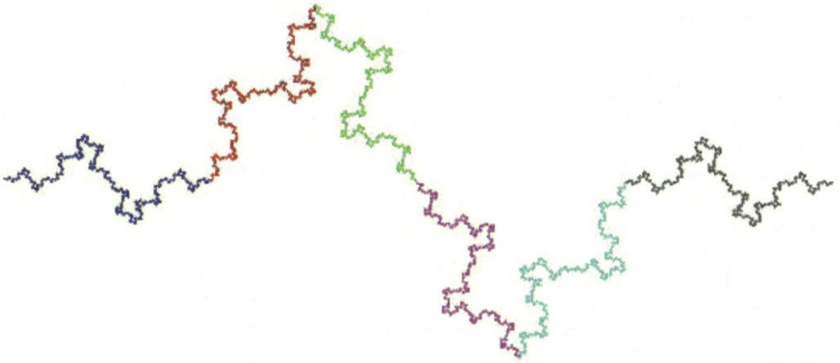

Figure 1.18. The approximating set $A_{(5)}$ for the fractal with generator G_2.

fractal A to consist of five similar copies of the form $\frac{1}{5}A$ (Figures 1.19, 1.20). The open quadrangle $(0,0)$, $(\frac{1}{3},\frac{1}{3})$, $(1,0)$, $(\frac{2}{3},-\frac{1}{3})$ provides the set for the open set condition, therefore the dimension of the corresponding fractal is again $\dim_S(A) = \frac{\log 5}{\log 3} \approx 1.46$.

Figure 1.19. The generator G_3. The grey lines delimit the parallelogram providing for the open set condition.

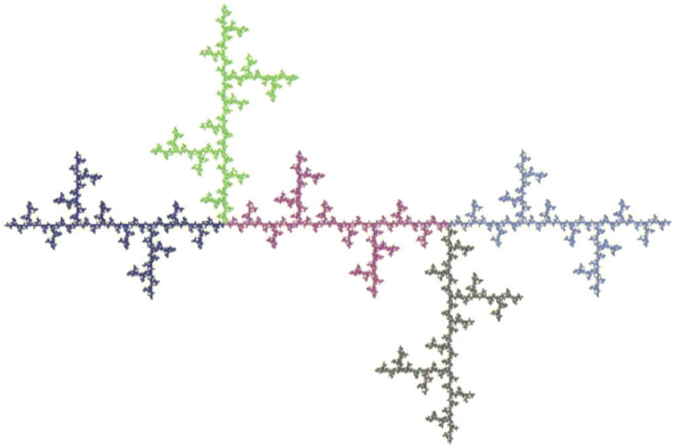

Figure 1.20. The approximating set $A_{(6)}$ for the fractal with generator G_3.

1.1.4.4 More modified Koch curves

A nice-looking fractal is obtained using the unit interval as initiator and a generator G_4 (Figure 1.21) consisting only of two segments joining the points

$$p_0 = (0,0) \qquad p_1 = (0.5, \tfrac{1}{2\sqrt{3}} \approx 0.2887) \qquad p_2 = (1,0).$$

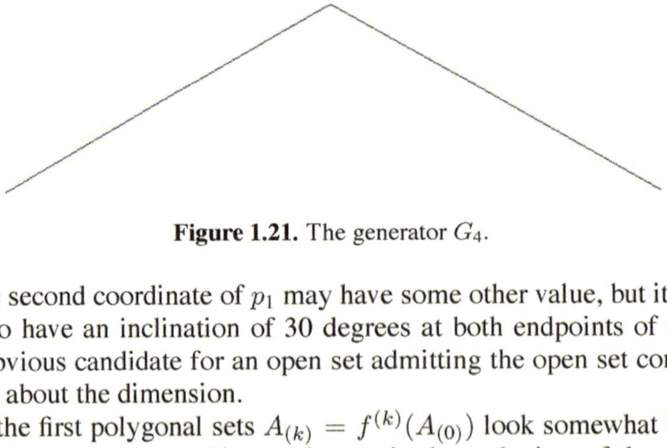

Figure 1.21. The generator G_4.

In fact, the second coordinate of p_1 may have some other value, but it is convenient at this point to have an inclination of 30 degrees at both endpoints of the generator. There is no obvious candidate for an open set admitting the open set condition, so we stop worrying about the dimension.

Although the first polygonal sets $A_{(k)} = f^{(k)}(A_{(0)})$ look somewhat harmless, like bathing caps, there are strange things going on in the endpoints of the segments constituting the sets $A_{(k)}$. Recall that, if p is such a point in $A_{(k_0)}$, it remains fixed under further applications of f. Let the left and right neighbouring points of p in $A_{(k)}$ be called $q_{(k)}$ and $\tilde{q}_{(k)}$ respectively and let us agree to paint everything to the left of p (with respect to the orientation of the curves $A_{(k)}$) blue and everything to the right of

p red. In contrast to $q_{(k+1)}p$ and $p\tilde{q}_{(k+1)}$ the blue segment $q_{(k)}q_{(k+1)}$ and the red segment $\tilde{q}_{(k+1)}\tilde{q}_{(k)}$ are not part of $A_{(k+1)}$ but for sake of supporting our intuition let these segments for the moment replace the part of $A_{(j)}$ $(j > k)$ between their endpoints. An application of f to $A_{(k)}$ produces a blue segment $q_{k+1}p$ by rotating the blue segment $q_{(k)}p$ clockwise about 30 degrees and shortening it by the factor $\frac{1}{\sqrt{3}}$. The same operation counterclockwise replaces $p\tilde{q}_{(k)}$ by $p\tilde{q}_{(k+1)}$. After at most six applications of f the points $q_{(k)}$ and $\tilde{q}_{(k)}$ coincide, causing the corresponding polygon $A_{(k)}$ to produce a loop (Figure 1.22). The blue segments $q_{(k)}q_{(k+1)}$ and the red segments $\tilde{q}_{(k+1)}\tilde{q}_{(k)}$ spiral in opposite direction about the point p and the same will be done by the fractal A. To give intuition the final blow: this happens at a dense set of points on the curve $A = \phi([0,1])$ (Figure 1.23).

A weird feature is added if we join two more points

$$p_3 = (0.8, 0) \qquad\qquad p_4 = (1, 0).$$

The last segment of the new generator G_5 (Figure 1.24) is now counted twice, as p_2p_3 and p_3p_4. Correspondingly, similar pictures of the generator appear above and below this segment. The effect is the apparition of loops, increasing to the right and making the fractal appear like a strange flower (Figure 1.25).

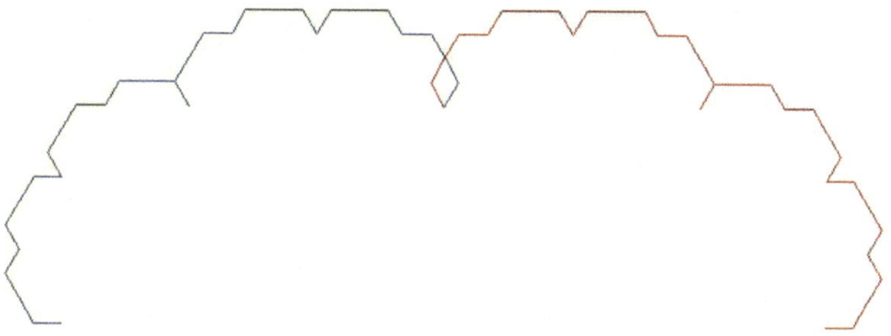

Figure 1.22. The approximating set $A_{(6)}$ for the fractal with generator G_4.

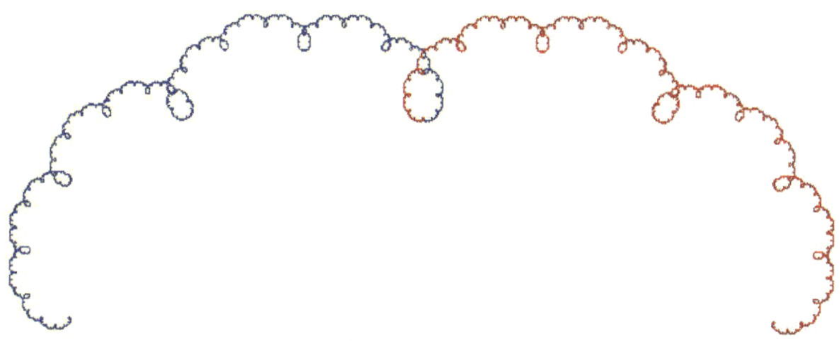

Figure 1.23. A closer approximation ($A_{(13)}$) of the fractal with generator G_4.

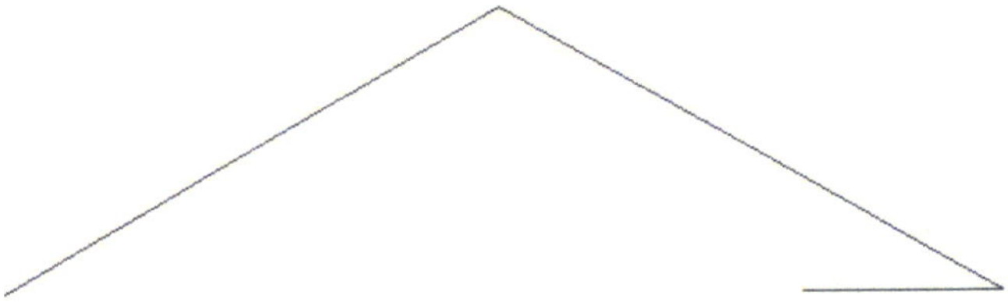

Figure 1.24. The generator G_5.

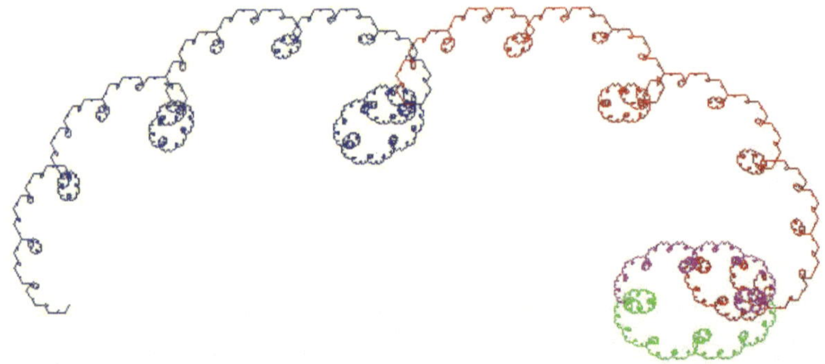

Figure 1.25. The approximating set $A_{(8)}$ for the fractal with generator G_5.

Loops also appear in a fractal as well behaved as produced by the generator G_6 (Figure 1.26) with three segments of equal length 0.5 and vertices

$$p_0 = (0,0) \qquad p_1 = (0.25, \tfrac{\sqrt{3}}{4} \approx 0.4330)$$
$$p_2 = (0.75, \tfrac{\sqrt{3}}{4}) \qquad p_3 = (1,0).$$

The first few sets $A_{(k)}$ seem to resemble a cauliflower, but eventually they settle down in form of a crystal with bubbling inclusions (Figure 1.27).

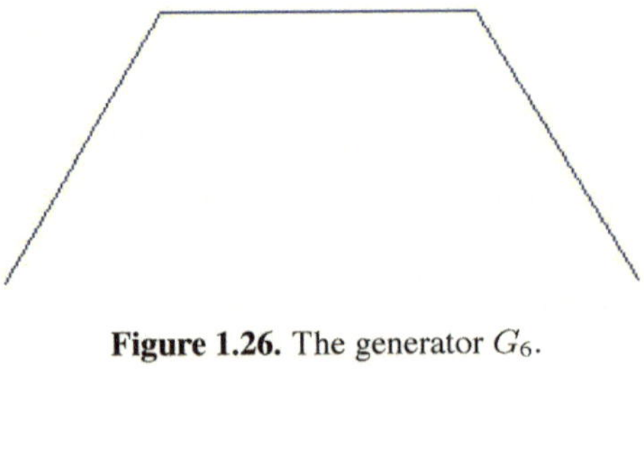

Figure 1.26. The generator G_6.

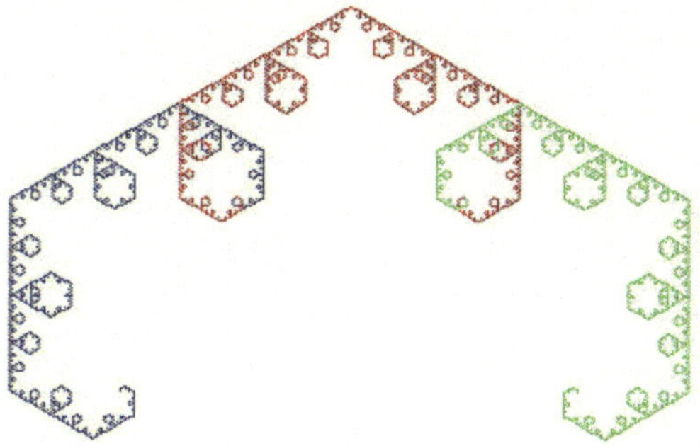

Figure 1.27. The approximating set $A_{(8)}$ for the fractal with generator G_6.

Finally, we can produce any weird fractal we wish by prescribing a suitably weird generator G_7. It suffices to let an error enter in the coordinates of the second example of Section 1.1.4.3 above (Figures 1.28, 1.29):

$$p_0 = (0,0) \qquad p_1 = (0.25,0) \qquad p_2 = (0.375, \tfrac{\sqrt{3}}{8} \approx 0.2165)$$
$$p_3 = (0.5,0) \qquad p_4 = (0.875, -\tfrac{\sqrt{3}}{8}) \qquad p_5 = (0.75,0)$$
$$p_6 = (1,0)$$

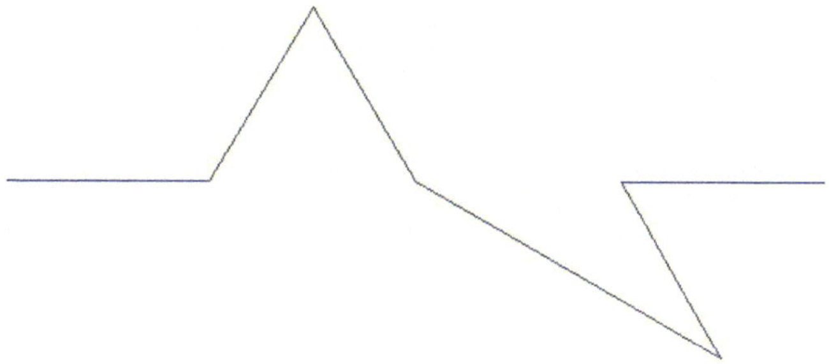

Figure 1.28. The generator G_7.

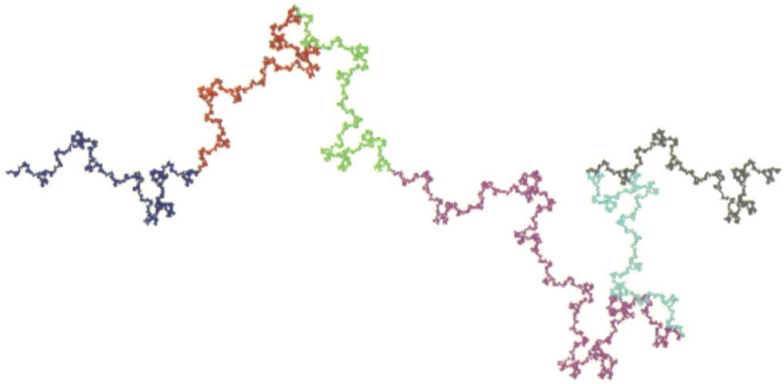

Figure 1.29. The approximating set $A_{(6)}$ for the fractal with generator G_7.

1.1.5 Space-filling curves

Surprising as it may sound, by a suitable choice of the generator it may happen that the resulting fractal, without losing its character as a curve, may pass through all points of an open set in the plane $\mathbb{R}^2$. Such a curve (here we only consider sets in $\mathbb{R}^2$) will be called *space-filling*.

1.1.5.1 The half square

Let the initiator $A_{(0)}$ again be the unit interval. Suppose the generator (Figure 1.30) connects the points

$$p_0 = (0,0) \qquad p_1 = (0.5,0) \qquad p_2 = (0.5,0.5)$$
$$p_3 = (0.5,0) \qquad p_4 = (1,0).$$

Considered as a curve, the polygon $A_{(1)}$ runs twice, in different directions, through the segment p_1p_2. The open triangle $p_0p_2p_4$ furnishes the open set condition since it

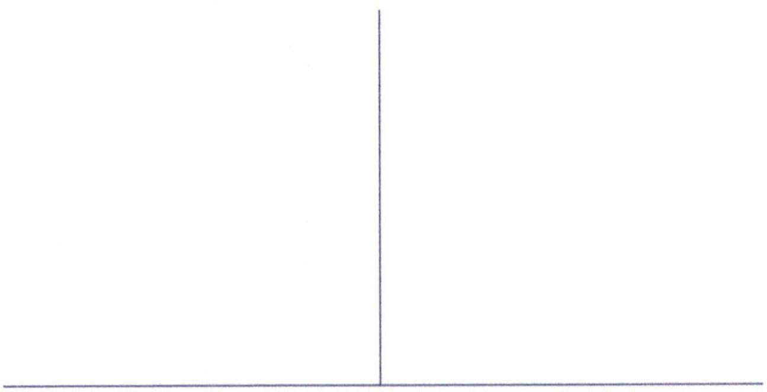

Figure 1.30. The generator of the half-square fractal.

contains its two open halves $p_0p_1p_2$ and $p_2p_3p_4$. Formula (1.7) gives $\dim_S(A) = \frac{\log 4}{\log 2} = 2$. Indeed, repeated application of the map f produces all segments of rectangular meshes of width $\frac{1}{2^k}$ parallel to the coordinate axes of the plane $\mathbb{R}^2$ which are contained in the closed triangle $p_0p_2p_4$ (Figure 1.31). Hence $\bigcup_{k=1}^{\infty} A_{(k)}$ is dense in this triangle and every point therein appears as a limit point, for some $t \in [0, 1]$, of the sequence $\phi_{(k)}(t)$ as described in Section 1.1.2. Figure 1.32 represents the graph of the function $\phi_{(7)}$ for $0 \le x \le 0.675$.

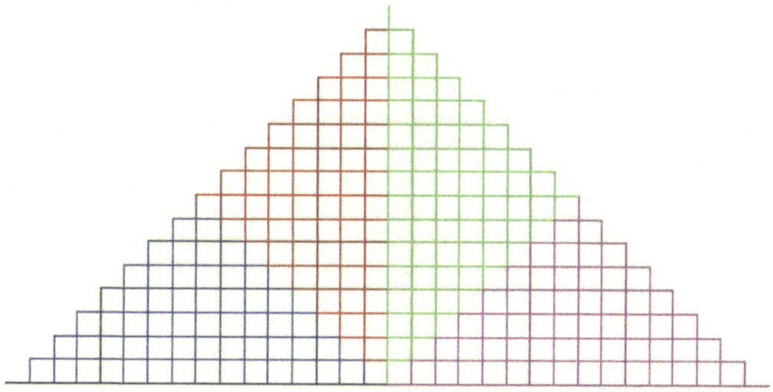

Figure 1.31. The approximating set $A_{(5)}$ for the half-square fractal.

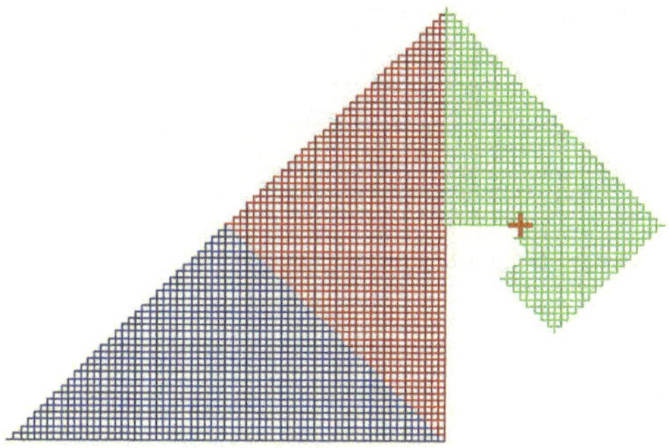

Figure 1.32. A closer approximation ($A_{(7)}$) of the half-square fractal; the graph of $\phi_{(7)}$ is drawn for $0 \le x \le 0.675$; the cross marks the point $\phi_{(7)}(0.675)$.

Although this means that indeed we can pass through every point of the whole triangle $p_0p_2p_4$ on a continuous curve, and although the construction is surprisingly simple, from an aesthetic point of view one might consider it not so satisfying that apart from the unit interval every segment in $A_{(k)}$ has to be passed through twice. Can we do better than that?

1.1.5.2 The crab

Take an equilateral triangle above the x-axis with basis $[0, \frac{1}{2}]$ and one below the x-axis with basis $[\frac{1}{2}, 1]$ and delete both bases (Figure 1.33). The resulting generator connects consecutively the points

$$p_0 = (0,0) \qquad p_1 = (0.25, \tfrac{\sqrt{3}}{4}) \qquad p_2 = (0.5,0)$$
$$p_3 = (0.75, -\tfrac{\sqrt{3}}{4}) \qquad p_4 = (1,0). \tag{1.8}$$

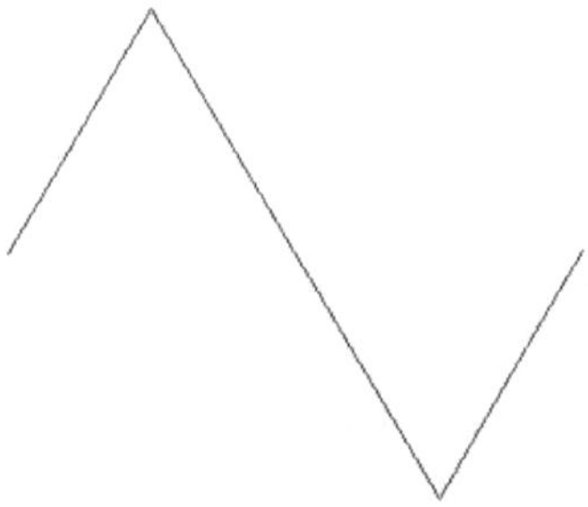

Figure 1.33. The generator $A_{(1)}$ of the crab.

Although there is no candidate set presenting itself for the open set condition, if for some good reason formula (1.7) would still be applicable, then for the corresponding fractal A it would give

$$\dim_S(A) = \tfrac{\log 4}{\log 2} = 2$$

which nourishes the suspicion that A might in some sense be genuinely a planar set.

In order to find out more about this, denote by g_0, g_1 and g_2 the three straight lines through the origin $(0,0)$ of the plane $\mathbb{R}^2$ with slopes 0 (the x-axis) and $\pm\sqrt{3}$. Consider the triangular mesh $\mathcal{T}_{(k)}$ generated by the three families of straight lines parallel to g_0, g_1 and g_2 and having, as pairwise distances, integral multiples of $\delta_k = \frac{\sqrt{3}}{2^{k+1}}$. Let us call the equilateral triangles of side length $\frac{1}{2^k}$ in $\mathcal{T}_{(k)}$ its *cells*. The generator is incident with $\mathcal{T}_{(1)}$, i.e. its four segments are sides of cells of $\mathcal{T}_{(1)}$. An application of f not only to the generator but to each segment of the mesh $\mathcal{T}_{(1)}$ maps $\mathcal{T}_{(1)}$ onto $\mathcal{T}_{(2)}$. It is readily seen that more generally we have $f(\mathcal{T}_{(k)}) = \mathcal{T}_{(k+1)}$. As a consequence, we get that $A_{(k)} = f^{(k)}(A_{(0)})$ is incident with $\mathcal{T}_{(k)}$.

Fix a cell $B_{(k)}$ in $\mathcal{T}_{(k)}$ (painted red) and surround it by the union $C_{(k)}$ of all outside adjacent cells of $\mathcal{T}_{(k)}$ (painted blue; in fact, $B_{(k)}$ is incident with $C_{(k)}$) (Figure 1.34).

Let us call two sets of segments *essentially disjoint* if the two corresponding unions of open component-segments are disjoint. An application of f to any segment outside of $C_{(k)}$ does not in any way afflict $B_{(k)}$. Inspection of the action of f on the segments constituting $\mathcal{T}_{(k)}$ we find:

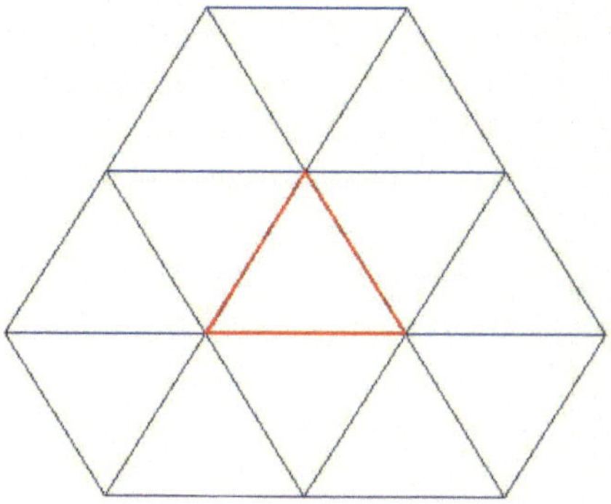

Figure 1.34. The cell $B_{(k)}$ (red) and the corresponding set $C_{(k)}$ (blue).

(a) the image of a segment does not depend on its orientation;

(b) images of disjoint open segments are essentially disjoint, images of adjacent segments produce adjacent segments in $\mathcal{T}_{(k+1)}$;

(c) two pairs of adjacent segments which do not traverse each other have images which do not traverse each other;

(d) the set $f(C_{(k)})$ contains all segments of all cells of $\mathcal{T}_{(k+1)}$ inside the image of the perimeter of $C_{(k)}$;

(e) the set $f(C_{(k)})$ contains the belt $C_{(k+1)}$ of all cells in $\mathcal{T}_{(k+1)}$ adjacent to and outside of $B_{(k)}$, i.e. its perimeter has distance δ_{k+1} from $B_{(k)}$;

(f) by induction, assertion (e) implies that the set $f^{(m)}(C_{(k)})$ contains all segments of all cells in $C_{(k+m)}$, i.e. its perimeter has distance δ_{k+m} from $B_{(k)}$;

(g) as a further consequence of (e), for every cell $B_{(k+1)}$ in $\mathcal{T}_{(k+1)}$ within $B_{(k)}$, a belt of all neighbouring cells in $\mathcal{T}_{(k+1)}$ is again contained in $f(C_{(k)})$;

(h) again by induction, all assertions (a)–(g) hold for every cell $B_{(j)}$ of $\mathcal{T}_{(j)}$ $(j > k)$ within $B_{(k)}$.

Several conclusions may be drawn from these facts:

(α) Since $A_{(1)}$, considered as a curve, passes through its segments only once and does not traverse itself, by (b) and (c) the same is true for $A_{(k)}$ for all $k > 1$.

(β) If, for some $k \in \mathbb{N}$, a set $C_{(k)}$ as described above is incident with $A_{(k)}$ (i.e. $A_{(k)}$, considered as a curve, runs through all segments of $C_{(k)}$), then by (d) and (f), for all $j > k$ the set $A_{(j)}$ contains the segments of all cells of $\mathcal{T}_{(j)}$ lying within the corresponding cell $B_{(k)}$ of $\mathcal{T}_{(k)}$.

(γ) In the situation described in (β) and for $j > k$, by (h) for every cell $B_{(j)}$ of $\mathcal{T}_{(j)}$ lying within $B_{(k)}$ a set $C_{(j)}$ of all cells of $\mathcal{T}_{(j)}$ adjacent to $B_{(j)}$ is again incident with $A_{(j)}$.

(δ) Recall from Section 1.1.2 how the fractal A may be considered as a curve $\phi([0,1])$ appearing as set of limit points of the approximating curves $A_{(k)} = \phi_{(k)}([0,1])$. Again in the situation described in (β), every point within $B_{(k)}$ is such a limit point (possibly for more than one parameter value t) and therefore must belong to A. Consequently, the curve A runs through every point of the set $B_{(k)}$, a set with a non-empty interior.

Back to our special generator (1.8): the set $A_{(5)} = f^{(5)}(A_{(0)})$ indeed (in several places in the area marked yellow) contains a set $C_{(5)}$ of cells in $\mathcal{T}_{(5)}$ as described above (Figure 1.35). Conclusion: the fractal A is space-filling. Still, this fact is not limited to the central cells of the just identified sets $C_{(5)}$ in $\mathcal{T}_{(5)}$. Since the resulting fractal is self-similar, it contains open sets in the neighbourhood of each of its points. As a continuous image of the unit interval it is closed. So it is the connected closure of a bounded open set in the plane $\mathbb{R}^2$. It is only due to blooming imagination that it is envisioned as a crab with scissors and legs marching sideways through the picture (Figure 1.36).

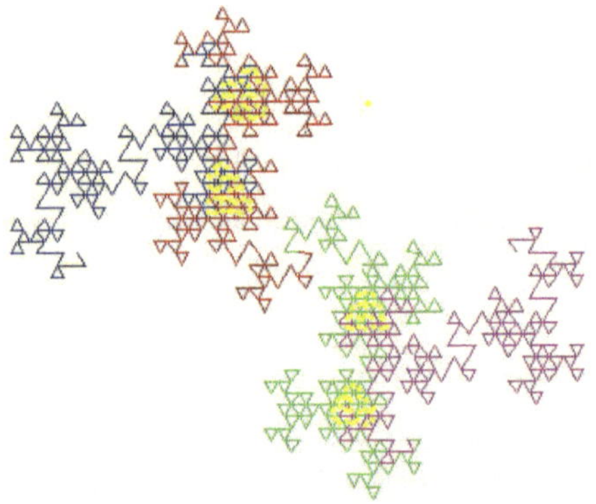

Figure 1.35. The approximating set $A_{(5)}$ for the crab. The areas indicated in yellow contain sets of type $C_{(5)}$ as described in the text.

The crab displays one more interesting feature related to what has been discussed in Section 1.1.4.4. Due to the positive slopes of the generator in the points $(0,0)$ and $(1,0)$, again it spirals about each vertex appearing in any set $A_{(k)}$ $(k \geq 1)$, but now the left (blue) and the right (red) spiral both turn counterclockwise without interfering with each other, the "blue" one leading into the vertex, and the "red" one leading out.

Figure 1.36. A closer approximation ($A_{(8)}$) of the crab fractal.

1.1.5.3 The Heighway-Harter dragon

A very simple generator consists of only two segments, the small side of a right-angled triangle with hypotenuse $[0,1]$ (Figure 1.37). Its vertices are

$$p_0 = (0,0) \qquad p_1 = (0.5, 0.5) \qquad p_2 = (1,0).$$

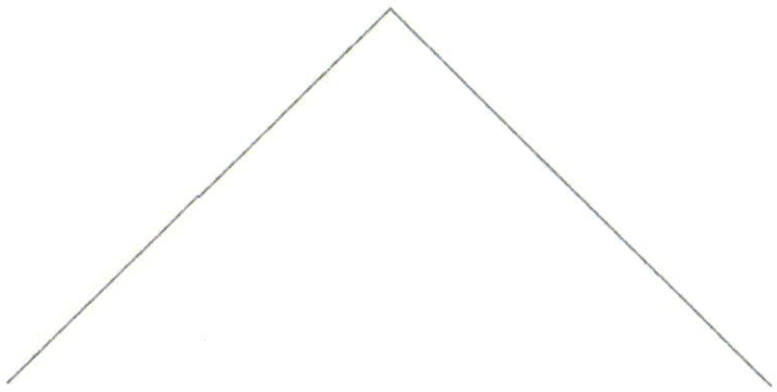

Figure 1.37. The generator $A_{(1)}$ of the dragon.

The initiator is again the interval $[0,1]$. A mapping f as defined so far produces a fractal A somewhat similar to the one generated in Section 1.1.4.4. It might be confusing that an application of formula (1.7) would seem to give $\dim_S(A) = \frac{\log 2}{\log \sqrt{2}} = 2$, but on second thought we have not verified the open set condition, and the loops appearing in A indicate that we shall not be able to do so if we try.

To complicate matters we shall now agree in the definition of the map f that it should replace each segment by the suitably diminished generator alternatingly to the right and to the left (cf. [Gardner, 1967], [Gardner, 1977], [Davis and Knuth, 1970]). Now it turns out that an application of $f^{(2k)}$ to $A_{(0)}$ produces a set $A_{(2k)}$ incident with the mesh $\mathcal{T}_{(k)}$ generated by the two families of lines parallel to the coordinate axes and distant from each other an integral multiple of $\frac{1}{2^k}$ (Figures 1.38–1.40).

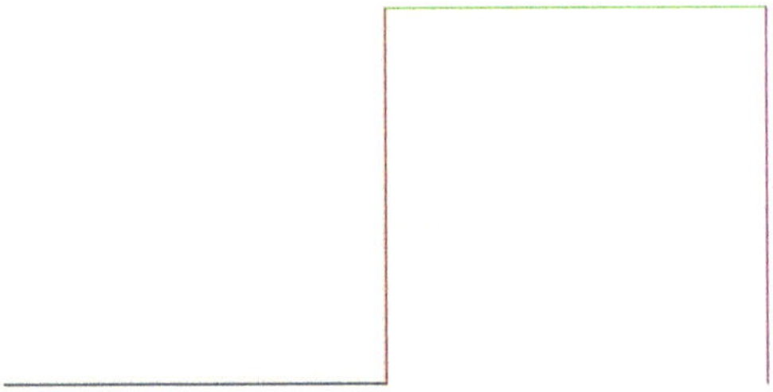

Figure 1.38. The approximating set $A_{(2)}$ for the dragon.

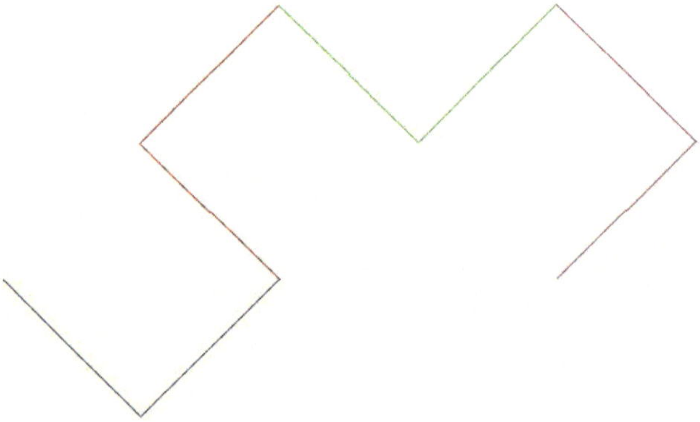

Figure 1.39. The approximation set $A_{(3)}$ of the dragon.

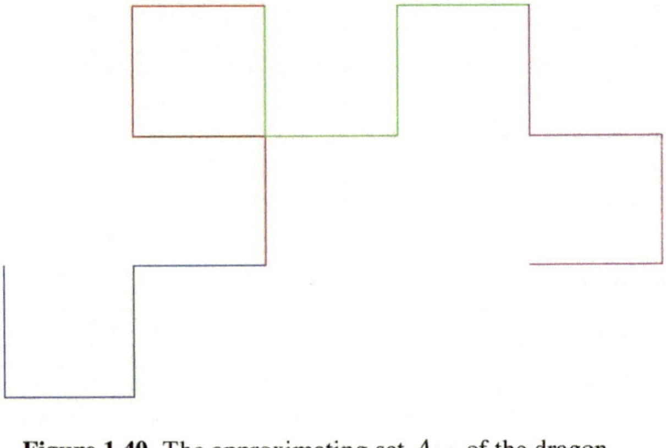

Figure 1.40. The approximating set $A_{(4)}$ of the dragon.

In fact an inspection similar to the one conducted in Section 1.1.5.2 reveals that indeed the resulting fractal A is space-filling: the set B is to be a replaced by a square of side length $\frac{1}{2^k}$, the set $C_{(2k)}$ by the set of all segments belonging to the square cells neighbouring B, but one has to be careful to mark the vertices of $A_{(2k)}$ where the generator $A_{(2)}$ of $f^{(2)}$ has to start replacing two adjacent segments in order to produce $A_{(2k+2)}$. The set $A_{(8)}$ contains a configuration of the form $C_{(8)}$ (Figure 1.41), and $A_{(2j)}$ $(j \geq 4)$ contains all segments of cells in $\mathcal{T}_{(j)}$ lying within the middle cell B. Therefore every point within the closed square B as a limit point of the sequence of sets $A_{(2k)}$ belongs to A. Again, for the reasons given in Section 1.1.5.2, the fractal A is the connected closed hull of its interior.

Figure 1.41. The approximating set $A_{(8)}$ for the dragon. Indicated in yellow is a set of type $C_{(8)}$ as described in the text.

The tail, the claws and fins, and the spiralling tongue might have inspired the choice of the name "dragon" for this fractal (Figure 1.42).

Figure 1.42. A closer approximation ($A_{(16)}$) of the dragon fractal. The white spots mark the screen pixels which are not yet visited by the piecewise linear curve furnishing the point set $A_{(16)}$.

1.1.5.4 The Peano curve

Although the full square (just as the half square) is hardly eligible for the name "fractal", there is a way, much more elegant than done in Section 1.1.5.1, by the process of repeatedly applying a suitable generator to the unit interval as initiator, to produce a curve A which passes through every point of this full square [Peano, 1890]. (A related construction avoiding multiple points is due to Hilbert, cf. [Hilbert, 1891].) This generator $A_{(1)}$ (Figure 1.43) connects the points

$$p_0 = (0,0) \qquad p_1 = (\tfrac{1}{3},0) \qquad p_2 = (\tfrac{1}{3},\tfrac{1}{3})$$
$$p_3 = (\tfrac{2}{3},\tfrac{1}{3}) \qquad p_4 = (\tfrac{2}{3},0) \qquad p_5 = (\tfrac{1}{3},0)\ (= p_1)$$
$$p_6 = (\tfrac{1}{3},-\tfrac{1}{3}) \qquad p_7 = (\tfrac{2}{3},-\tfrac{1}{3}) \qquad p_8 = (\tfrac{2}{3},0)\ (= p_4)$$
$$p_9 = (1,0).$$

It is readily seen that the open square Q with the diagonal p_0p_9 furnishes the open set condition and that for the resulting set A, formula (1.7) gives $\dim_S(A) = \frac{\log 9}{\log 3} = 2$. In fact, it is also readily seen that every point of the closure of Q is a limit point of the sequence of sets $A_{(k)}$ (Figure 1.44).

So the method employed in the study of the Koch curve should produce a continuous function ϕ on the unit interval passing through every point of the closed square Q. Let us try to construct it explicitly.

In order to simplify the computation let us turn the two coordinate axes clockwise about $\frac{\pi}{4}$ so that the y-axis contains the point p_2, and redefine the side of Q to have unit

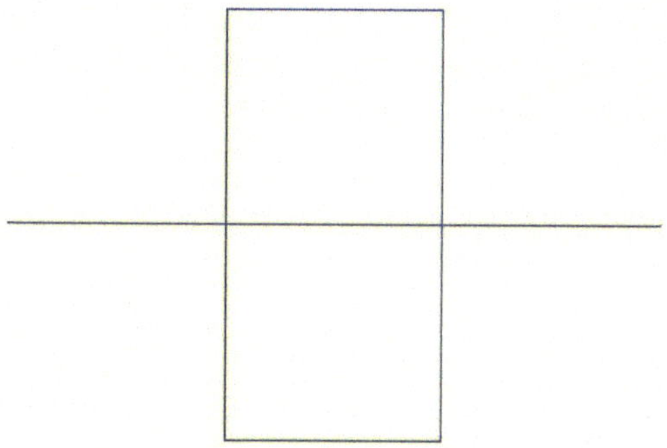

Figure 1.43. The generator $A_{(1)}$ for the PEANO curve.

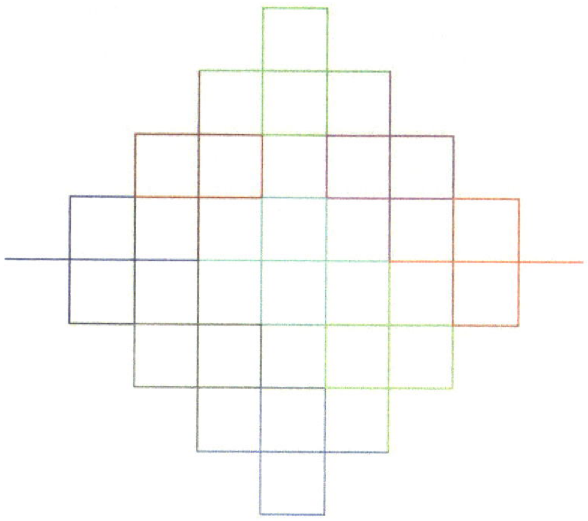

Figure 1.44. The approximating set $A_{(2)}$ of the PEANO curve.

length. As in Section 1.1.1 we define the piecewise linear mapping $\phi_{(1)}$ on $[0,1]$ to map each of the intervals $[\frac{j}{9}, \frac{j+1}{9}]$ linearly onto the segment A_j of $A_{(1)}$ $(0 \le j \le 8)$. As a curve, $\phi_{(1)}([0,1])$ passes through the diagonals of consecutive sub-squares Q_j with side length $\frac{1}{3}$. Due to the varying inclination of segments in $A_{(1)}$ we attach to each sub-square Q_j an integer m_j (mod 4) as inclination coefficient, which indicates that the segment A_j serving as diagonal of Q_j is turned clockwise about a hook of $m_j \frac{\pi}{2}$ with respect to the position of p_0p_9. Thus we have

$$\begin{array}{lll} m_0 = 0 & m_1 = 3 & m_2 = 0 \\ m_3 = 1 & m_4 = 2 & m_5 = 1 \\ m_6 = 0 & m_7 = 3 & m_8 = 0. \end{array}$$

Let the point $p = (x, y)$ in the closed square Q be given and consider the (if possible, finite) triadic decompositions

$$x = \sum_{i=1}^{\infty} \frac{x_i}{3^i} = 0.x_1x_2\ldots x_i\ldots \qquad (0 \le x_i \le 2),$$

$$y = \sum_{i=1}^{\infty} \frac{y_i}{3^i} = 0.y_1y_2\ldots y_i\ldots \qquad (0 \le y_i \le 2).$$

The digit x_1 determines the column of three sub-squares Q_j in which p lies, the digit y_1 determines the row of three sub-squares in which p lies. In fact, the subsegment A_{j_1} of $A_{(1)}$ forming the diagonal of the sub-square Q_{j_1} containing the point p (if p belongs to more than one sub-square Q_j, then j_1 is taken to be minimal) has consecutive number

$$j_1 = \begin{cases} y_1 & \text{if } x_1 = 0, \\ 5 - y_1 & \text{if } x_1 = 1, \\ 6 + y_1 & \text{if } x_1 = 2. \end{cases}$$

On $A_{(1)} = \phi_{(1)}([0,1])$ and within Q_{j_1} we approximate the point p by the point $p_{(1)} = \phi_{(1)}(\frac{j_1}{9})$, the initial point of the segment A_{j_1}.

An application of f to $A_{(1)}$ replaces the diagonal A_{j_1} of Q_{j_1} with a copy similar to $A_{(1)}$ but diminished by a factor $\frac{1}{3}$, producing inside of Q_{j_1} a string of segments $A_{j_1,j}$ $(0 \le j \le 8)$. Again this corresponds to a subdivision of Q_{j_1} into nine sub-squares $Q_{j_1,j}$ of side length $\frac{1}{3^2}$ each. As before, the mapping $\phi_{(2)}$ is defined linearly on every interval $[\frac{j_1}{9} + \frac{j}{9^2}, \frac{j_1}{9} + \frac{j+1}{9^2}]$ onto $A_{j_1,j}$. Again we want to approximate the point $p \in Q_{j_1}$ by the initial point of the segment A_{j_1,j_2} which is the diagonal of the sub-square Q_{j_1,j_2} containing p (if p is contained in more than one of these sub-squares we take the one with minimal index j_2). We proceed as before but in order to find j_2 in terms of x and y we have to take into account the inclination of Q_{j_1} indicated by m_{j_1}. Inside of Q_{j_1} the index j_2 of the sub-square Q_{j_1,j_2} containing p will again be determined by the triadic digits x_2 and y_2, but the role played before by x_1 and y_1 will now be played in case of

$m_{j_1} = 0$	by	$\tilde{x}_2 = x_2$	and	$\tilde{y}_2 = y_2,$
$m_{j_1} = 1$	by	$\tilde{x}_2 = 2 - y_2$	and	$\tilde{y}_2 = x_2,$
$m_{j_1} = 2$	by	$\tilde{x}_2 = 2 - x_2$	and	$\tilde{y}_2 = 2 - y_2,$
$m_{j_1} = 3$	by	$\tilde{x}_2 = y_2$	and	$\tilde{y}_2 = 2 - x_2.$

Correspondingly,

$$j_2 = \begin{cases} \tilde{y}_2 & \text{if } \tilde{x}_2 = 0, \\ 5 - \tilde{y}_2 & \text{if } \tilde{x}_2 = 1, \\ 6 + \tilde{y}_2 & \text{if } \tilde{x}_2 = 2. \end{cases}$$

On $A_{(2)} = \phi_{(2)}([0, 1])$ the point approximating p is taken to be $p_{(2)} = \phi_{(2)}(\frac{j_1}{9} + \frac{j_2}{9^2})$, the initial point of the segment A_{j_1,j_2}. The distance between p and $p_{(2)}$ which both lie in Q_{j_1,j_2} is at most $\frac{\sqrt{2}}{3^2}$. Note that the inclination coefficient of Q_{j_1,j_2} is $m_{j_1} + m_{j_2} \pmod 4$.

Proceeding by induction we obtain a sequence of continuous piecewise linear mappings $\phi_{(k)}$ mapping the intervals $[\sum_{i=1}^{k} \frac{n_i}{9^i}, \sum_{i=1}^{k} \frac{n_i}{9^i} + \frac{1}{9^k}]$ $(0 \leq n_i \leq 8)$ linearly onto $A_{n_1,\ldots,n_k}$; furthermore a sequence of segments $A_{j_1,\ldots,j_k}$ with initial points $p_{(k)} = \phi_{(k)}(\sum_{i=1}^{k} \frac{j_i}{9^i})$ distant from p at most $\frac{\sqrt{2}}{3^k}$, and finally a sequence of corresponding inclination coefficients $\sum_{i=1}^{k} m_{j_i} \pmod 4$. For $l > k$ and $t \in [\sum_{i=1}^{k} \frac{n_i}{9^i}, \sum_{i=1}^{k} \frac{n_i}{9^i} + \frac{1}{9^k}]$ the points $\phi_{(k)}(t)$ and $\phi_{(l)}(t)$ are at most a distance of $\frac{\sqrt{2}}{3^k}$ apart since both lie in $Q_{n_1,\ldots,n_k}$. Consequently, the sequence $\{\phi_{(k)}\}_{k=1}^{\infty}$ converges uniformly to a continuous mapping ϕ and $\phi(\sum_{i=1}^{\infty} \frac{j_i}{9^i}) = \lim_{k\to\infty} \phi(\sum_{i=1}^{k} \frac{j_i}{9^i}) = \lim_{k\to\infty} \phi_{(k)}(\sum_{i=1}^{k} \frac{j_i}{9^i}) = p$. In other words, we have established that the continuous curve $\phi([0, 1])$ indeed passes through every given point $p \in Q$.

Figure 1.45. A closer approximation of the PEANO curve. The graph of $\phi_{(5)}$ is drawn for $0 \le x \le 0.833$. The cross indicates the point $\phi_{(5)}(0.833)$. The argument $x = 0.833$ has 9-adic expansion $0.7442233153\ldots$.

Figure 1.46. A tessellation of the unit square by sub-squares Q_{j_1,j_2} $(0 \le j_i \le 8)$ of side length $\frac{1}{3^2}$ in the order in which they are passed through by the PEANO curve. In this order the sub-squares are painted consecutively in the colours blue, red, green, magenta, cyan, dark grey, light blue, light green, light red. The tessellation starts in the left corner, going up first in north-east direction, and ends in the right corner.

1.1.6 Short fractal curves

Although the KOCH curve is the image under a continuous mapping of the unit interval into $\mathbb{R}^2$ – so we may pass through it in finite time – it has infinite length. The same is valid for the fractal curves we have discussed thereafter, not withstanding the fact that we have not always stated this explicitly. So rather naturally the question came up in a course: "Are there also fractal curves of finite length?" There is a sloppy answer to it: "Yes, the CANTOR set!" But this is not quite satisfying since here we meet the other extreme: if "length" should have some meaning for the CANTOR set, it would have to be measured by LEBESGUE measure which gives it the value zero. So the next question asks whether there is a fractal curve – whatever this may mean – which has in some natural sense a finite but positive measure?

1.1.6.1 A thick CANTOR set

Why has the CANTOR set got LEBESGUE measure zero? Because at every stage $A_{(k)}$ of its construction we have deleted disjoint subintervals from $A_{(k)}$ of total length $\frac{1}{3}\mathcal{L}(A_{(k)})$, thereby leaving a set $A_{(k+1)}$ of LEBESGUE measure $\mathcal{L}(A_{(k+1)}) = (\frac{2}{3})^{k+1}$. As $k \to \infty$ this gives $\mathcal{L}(A) = \mathcal{L}(\bigcap_{k=1}^{\infty} A_{(k)}) = \lim_{k\to\infty} \mathcal{L}(A_{(k)}) = 0$. It seems prudent to delete in a more cautious way. Let any $\varepsilon < \frac{1}{3}$ be given and start out again with the closed unit interval $A_{(0)}$. At the k-th stage ($k \geq 1$), delete from every component interval of $A_{(k-1)}$ (there are 2^{k-1} of them) a centrally positioned open subinterval of length $\frac{\varepsilon}{2^{2k-1}}$. This leaves a closed set $A_{(k)}$ with LEBESGUE measure $1 - \sum_{i=1}^{k} 2^{i-1} \cdot \frac{\varepsilon}{2^{2i-1}} = 1 - \varepsilon \sum_{i=1}^{k} \frac{1}{2^i} = 1 - \varepsilon(1 - \frac{1}{2^k})$. The set $A = \bigcap_{k=1}^{\infty} A_{(k)}$ is closed, totally disconnected (thereby topologically zero-dimensional), perfect, nowhere dense, topologically homeomorphic with the CANTOR set, and has LEBESGUE measure $1 - \varepsilon > 0$, as close to 1 as desired.

Due to the fact that the total length of the deleted subintervals is not proportional to the total length of the set from which they are deleted, we do not have exact self-similarity within A. So as to dimension we better wait until later.

1.1.6.2 A self-similar fractal curve with positive finite "length"

Self-similarity of a fractal curve is certainly present if we use the method of initiator and generator. Still, in the cases considered so far, a continuous piecewise linear generator different from the unit interval $A_{(0)}$ but connecting its end points is bound to have length larger than 1. With every application of f to $A_{(k)}$ ($k \geq 0$) the length of $A_{(k)}$ is multiplied with the length of the generator. Consequently, the length of $A_{(k)}$ for $k \to \infty$ eventually exceeds all bounds, and the length of the limiting set A, which cannot be smaller, must be infinite.

If we want to keep the length of $A_{(k)}$ bounded, our only chance is to cut the unit interval into finitely many pieces and to place them disjointly from each other. This will destroy continuity in the resulting fractal, but we have already made peace with this fact when dealing with the CANTOR set.

In order to keep things simple, consider the generator $A_{(1)}$ (Figure 1.47) consisting of two segments p_0p_1 and p_2p_3 with

$$p_0 = (0,0) \qquad p_1 = (0.5,0)$$
$$p_2 = (1,0.5) \qquad p_3 = (1,1).$$

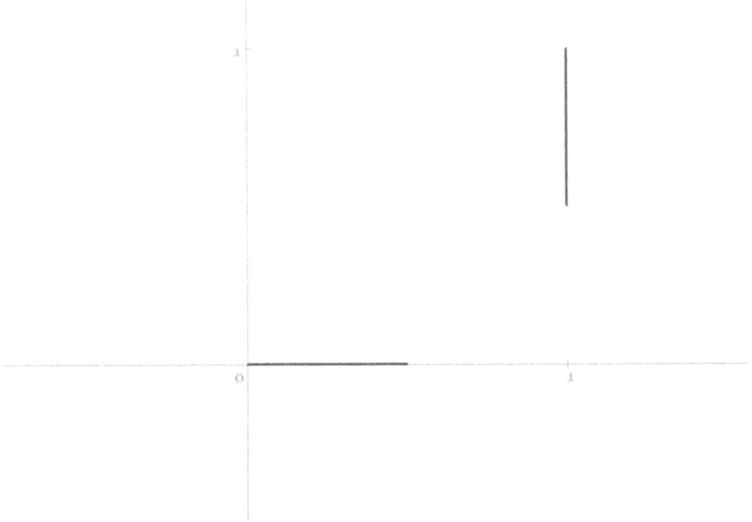

Figure 1.47. The generator $A_{(1)}$ for the "short fractal curve".

Again the application of the mapping f consists in replacing every subsegment of the set $A_{(k-1)}$ with a suitably diminished copy of $A_{(1)}$. It is readily seen (e.g. by induction) that $A_{(k)}$ consists of 2^k subsegments of length $\frac{1}{2^k}$ each, so the total length of the subsegments of $A_{(k)}$ always remains 1. Unfortunately this fact alone does not allow us to conclude that the limiting set A (Figure 1.48) in some sense also has to have length 1. First of all, the points of A lie scattered around in the unit square and it is not clear how to attribute a length to their union. Secondly, even in the continuous case of a staircase-curve from p_0 to p_3 with steps of width and height $\frac{1}{2^k}$, having total length 2 for every $k \in \mathbb{N}$, the limiting curve would be the diagonal p_0p_3 with length $\sqrt{2}$.

As an encouraging fact, the open unit square furnishes a set needed for the open set condition, which justifies the computation $\dim_S(A) = \frac{\log 2}{\log 2} = 1$. But, as explained above, the computation of a length by means of LEBESGUE measure breaks down. Perhaps we can simulate the computation of LEBESGUE measure of a subset of the real line (by means of coverings by open intervals of arbitrarily small length) and find coverings of A by means of sets of arbitrarily small diameter and bounded sum of all diameters. Indeed, the set $A_{(1)}$ is contained in the union $P_{(1)}$ of two closed sub-squares Q_j $(0 \le j \le 1))$ of the unit square Q sharing its diagonal and having side length $\frac{1}{2}$ each. Similarly, $A_{(2)}$ is contained in the union $P_{(2)}$ of four closed sub-squares Q_{j_1,j_2} $(0 \le j_i \le 1)$ of Q of side length $\frac{1}{4}$ each, contained in $P_{(1)}$. By induction we

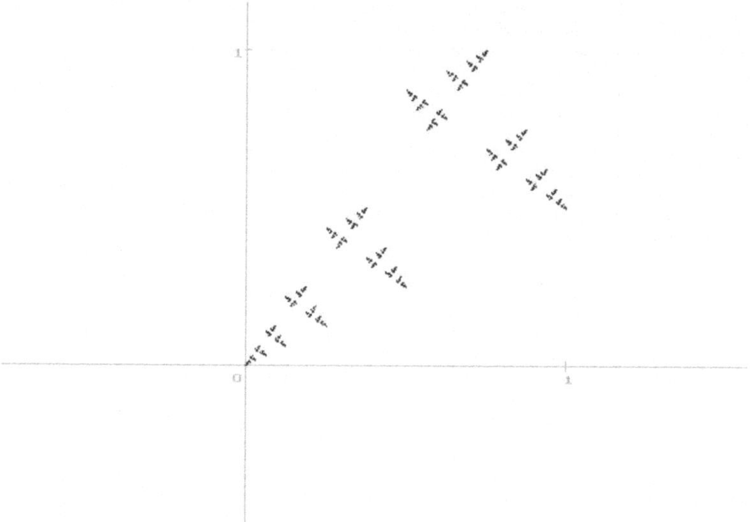

Figure 1.48. A closer approximation $(A_{(9)})$ of the "short fractal curve".

find that $A_{(k)}$ is contained in the union $P_{(k)}$ of 2^k closed sub-squares $Q_{j_1,\dots,j_k}$ of side length $\frac{1}{2^k}$ each and that $P_{(1)} \supset P_{(2)} \supset \cdots \supset P_{(k)}$. The set A consists of limit points of the sequence of closed sets $\{A_{(k)}\}_{k=1}^{\infty}$. All of these limits points are contained in $\bigcap_{k=1}^{\infty} P_{(k)}$ and therefore, for each $k \in \mathbb{N}$, in $P_{(k)}$. In other words, A is covered by the 2^k sub-squares $Q_{j_1,\dots,j_k}$ mentioned above. The sum of their diameters amounts to $\sqrt{2}$.

We have just computed an upper bound for the one-dimensional HAUSDORFF *measure* of A which is defined by

$$\begin{aligned} \mathcal{H}^1(A) &:= \sup_{\delta>0} \inf \Big\{ \sum_{i=1}^{\infty} \operatorname{diam}(C_i) : A \subset \bigcup_{i=1}^{\infty} C_i,\ \operatorname{diam}(C_i) \le \delta\ (i \in \mathbb{N}) \Big\} \\ &= \lim_{\delta \to 0} \inf \Big\{ \sum_{i=1}^{\infty} \operatorname{diam}(C_i) : A \subset \bigcup_{i=1}^{\infty} C_i,\ \operatorname{diam}(C_i) \le \delta\ (i \in \mathbb{N}) \Big\}. \end{aligned} \tag{1.9}$$

Recall that $\operatorname{diam}(C) := \sup\{|x - y| : x \in C, y \in C\}$. As will be established later, on the real line the one-dimensional HAUSDORFF measure coincides with LEBESGUE measure. So for a legitimate extension $\mathcal{H}^1$ of the concepts of "length" and "one-dimensional LEBESGUE measure" we have found that $\mathcal{H}^1(A) \le \sqrt{2}$.

It would be disappointing if it turned out that the HAUSDORFF measure of A would in fact be zero. In order to obtain a lower bound for $\mathcal{H}^1(A)$, observe that each of the projections $p_x(A_{(k)})$ and $p_y(A_{(k)})$ of the set $A_{(k)}$ upon the x-axis and y-axis respectively consists of disjoint intervals of length at most $\frac{1}{2^k}$ each (some of them just consist of one single point), spaced at most a $\frac{1}{2^k}$ from the neighbouring one. Again this may be proved by induction: the assertion is true for $k = 1$; assuming its truth for some k we note that $A_{(k+1)}$ is obtained by placing a similar copy half the size of $A_{(k)}$ into the lower left sub-square Q_1 and an equal copy, turned about an angle of $\frac{\pi}{2}$ counterclockwise, into the upper right sub-square Q_4. Together their projections supply the wanted family of intervals on the x-axis and y-axis. The set of limit points of the sequences of sets $p_x(A_{(k)})$ and $p_y(A_{(k)})$ as $k \to \infty$ therefore consists of all points of the unit interval on the x-axis and y-axis respectively.

As mappings of $\mathbb{R}^2$ upon the x-axis and y-axis respectively, the projections p_x and p_y are continuous (the following statements, formulated for p_x, hold for p_y as well). For any two points p and q in the plane, p_x even satisfies the inequality

$$|p_x(p) - p_x(q)| \leq |p - q|.$$

If $A \subset \bigcup_{i=1}^{\infty} C_i$ and $\operatorname{diam}(C_i) \leq \delta$ for all $i \in \mathbb{N}$, then $p_x(A) \subset \bigcup_{i=1}^{\infty} p_x(C_i)$, $\operatorname{diam}(p_x(C_i)) \leq \delta$ and $\sum_{i=1}^{\infty} \operatorname{diam}(C_i) \geq \sum_{i=1}^{\infty} \operatorname{diam}(p_x(C_i))$. For any set $A \in \mathbb{R}^2$ this implies

$$\begin{aligned}
&\inf\Big\{\sum_{i=1}^{\infty} \operatorname{diam}(C_i) : A \subset \bigcup_{i=1}^{\infty} C_i,\ \operatorname{diam}(C_i) \leq \delta\ (i \in \mathbb{N})\Big\} \\
&\qquad \geq \inf\Big\{\sum_{i=1}^{\infty} \operatorname{diam}(C_i') : p_x(A) \subset \bigcup_{i=1}^{\infty} C_i',\ \operatorname{diam}(C_i') \leq \delta\ (i \in \mathbb{N})\Big\}, \\
&\sup_{\delta>0}\inf\Big\{\sum_{i=1}^{\infty} \operatorname{diam}(C_i) : A \subset \bigcup_{i=1}^{\infty} C_i,\ \operatorname{diam}(C_i) \leq \delta\ (i \in \mathbb{N})\Big\} \\
&\qquad \geq \sup_{\delta>0}\inf\Big\{\sum_{i=1}^{\infty} \operatorname{diam}(C_i') : p_x(A) \subset \bigcup_{i=1}^{\infty} C_i',\ \operatorname{diam}(C_i') \leq \delta\ (i \in \mathbb{N})\Big\}, \\
&\mathcal{H}^1(A) \geq \mathcal{H}^1(p_x(A)). \qquad (1.10)
\end{aligned}$$

The sets $A_{(k)}$ converge for $k \to \infty$ pointwise to the set A. Because of the continuity of p_x we get $p_x(A) = \lim_{k\to\infty} p_x(A_{(k)})$ i.e. the set of limit points of the sets $p_k(A_{(k)})$ as $k \to \infty$. We already know this set to be the unit interval. Consequently we obtain $\mathcal{H}^1(A) \geq 1$.

For anyone already familiar with the concept of HAUSDORFF dimension, our estimates again imply $\dim_H(A) = 1$ as an immediate consequence of Definition 1.3.5.

1.1.6.3 The CANTOR staircase

Deletion of the middle third pieces in the construction of the CANTOR set does not mean that these pieces are useless. Recall that in the construction of the CANTOR set the $(k-1)$-th configuration $A_{(k-1)}$ consists of $2^{(k-1)}$ disjoint intervals $A_{j_1,\dots,j_{k-1}}$ ($j_i \in \{0, 2\}$) of length $\frac{1}{3^{k-1}}$ each. The next configuration $A_{(k)}$ is obtained from $A_{(k-1)}$ by deleting in each one of those intervals the open middle third of length $\frac{1}{3^k}$. We shall denote these deleted intervals in their consecutive order $B_{(k),j}$ $(1 \leq j \leq 2^{(k-1)})$. We now start with lifting $B_{(1),1}$ within the unit square to height $y = \frac{1}{2}$. Connect its left endpoint $b_{(1),1,0}$ by a line segment with $(0, 0)$ and its right endpoint $b_{(1),1,1}$ by a line segment with $(1,1)$. We shall denote this configuration of three segments $C_{(1)}$ (Figure 1.49).

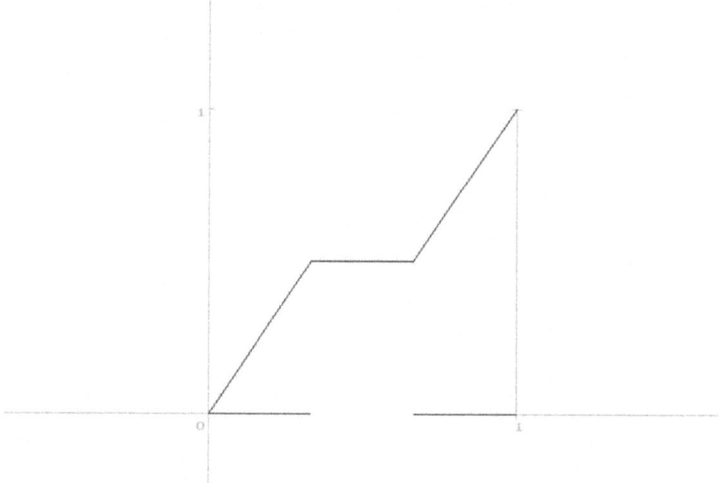

Figure 1.49. The approximating set $C_{(1)}$ for the CANTOR staircase.

In order to construct the next configuration $C_{(2)}$ we now replace the first and the last segment by the following configurations of three segments each: lift $B_{(2),1}$ to height $y = \frac{1}{2^2}$ and connect its left endpoint $b_{(2),1,0}$ as before with $(0,0)$, similarly its right endpoint $b_{(2),1,1}$ with $b_{(1),1,0}$; lift $B_{(2),2}$ to height $y = \frac{3}{2^2}$ and connect its left endpoint $b_{(2),2,0}$ with $b_{(1),1,1}$ and its right endpoint $b_{(2),2,1}$ with $(1,1)$. The configurations, obtained by these replacements, like upon $C_{(1)}$ but are no similar copies thereof. Their widths are $\frac{1}{3}$ of the width of $C_{(1)}$, but their heights are $\frac{1}{2}$ of the height of $C_{(1)}$. While the slope of the two connecting segments in $C_{(1)}$ was $\frac{3}{2}$, the slope of the four connecting segments in $C_{(2)}$ is $(\frac{3}{2})^2$.

Proceeding by induction, at the k-th step we lift every interval $B_{(k),j}$ $(1 \leq j \leq 2^{k-1})$ to height $y = \frac{2j-1}{2^k}$ and connect its endpoints with the endpoints of the neighbouring lifted x-parallel intervals in $C_{(k-1)}$, thereby arriving at the configuration $C_{(k)}$. It consists of $2^k - 1$ x-parallel intervals of total length $1 - (\frac{2}{3})^k$ and of 2^k segments connecting points with horizontal distance $\frac{1}{3^k}$ and vertical distance $\frac{1}{2^k}$; their slope therefore is $(\frac{3}{2})^k$ and their total length is $2^k\sqrt{\frac{1}{3^{2k}} + \frac{1}{2^{2k}}}$. The total length of the configuration $C_{(k)}$ is $1 - (\frac{2}{3})^k + \sqrt{(\frac{2}{3})^{2k} + 1}$ (Figure 1.50).

It is readily seen that $C_{(k)}$ is the graph of a continuous piecewise linear function ϕ_k on $[0,1]$ which, as $k \to \infty$, converges uniformly to a continuous function ϕ. The graph of ϕ has a horizontal tangent at each point of $\bigcup_{k=1}^{\infty} \bigcup_{j=1}^{2^{k-1}} B_{(k),j}$, an open set of LEBESGUE measure 1. On its complement, the CANTOR set, it has vertical tangents. The function ϕ, called LEBESGUE*'s singular function* [Lebesgue, 1905], is almost everywhere differentiable with vanishing derivative, but it is not the indefinite

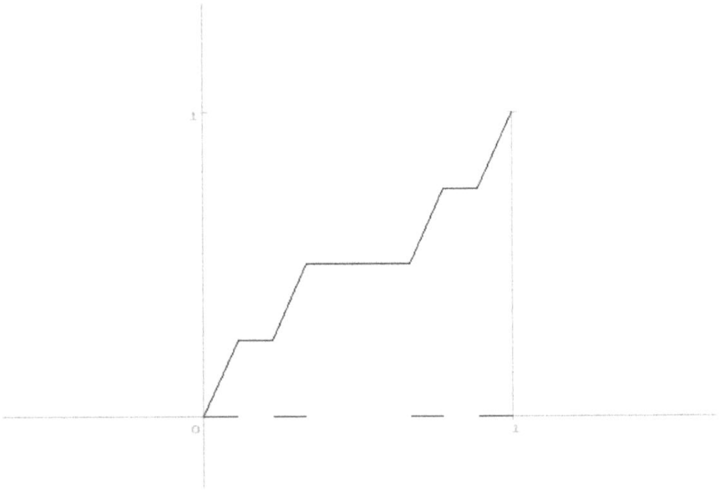

Figure 1.50. The approximating set $C_{(2)}$ for the CANTOR staircase.

integral of its derivative (which would be a constant function). The strange feature of it is that while its tangents are almost everywhere horizontal, it increases, and actually increases only on the CANTOR set, a set of LEBESGUE measure zero, from zero to one. Its graph C is frequently called CANTOR*'s staircase* and, probably because of the mentioned weird feature, sometimes *devil's staircase* (Figure 1.51).

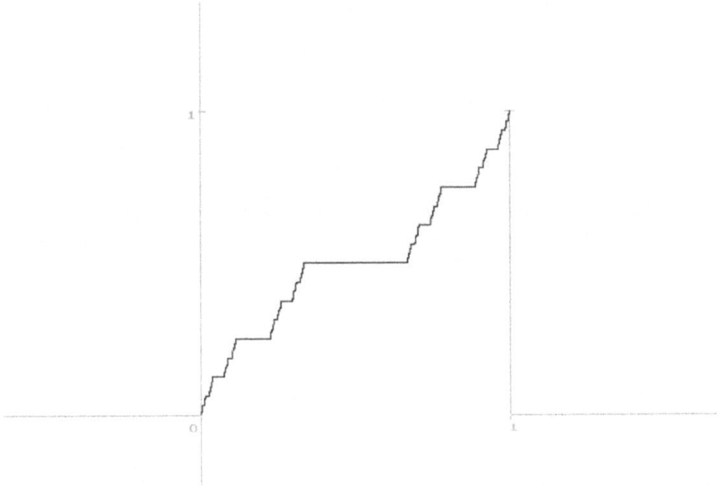

Figure 1.51. A closer approximation ($C_{(8)}$) of the CANTOR staircase.

It is tame with respect to its length. As sequences of segments connecting points of C, the configurations $C_{(k)}$ are perfectly legitimate means to compute, in the limit as $k \to \infty$, the length of C as a curve. From what has been computed above this limit is

$\lim_{k\to\infty}\left(1-(\frac{2}{3})^k+\sqrt{(\frac{2}{3})^{2k}+1}\right)=2.$

Although CANTOR's staircase tries hard to look self-similar, it is not, due to the reasons indicated above. This might be an indication that one should not be too strict with self-similarity when dealing with fractals. In fact we shall meet more general concepts in the second chapter.

1.1.7 Higher dimensional CANTOR sets

The construction of the CANTOR set started with a one-dimensional line segment and produced a set with a smaller but still positive dimension. We shall now try to see whether this procedure may be modified so as to be applicable also to initial sets of greater dimension. If we succeed with some examples in two- or three-dimensional space then we should have no trouble in trusting that similar fractals may be constructed in spaces of arbitrary dimension.

1.1.7.1 The SIERPINSKI triangle

Take an equilateral triangle $A_{(0)}$ with side length 1. In fact any other triangle would do as well, but for symmetry reasons such a setup gives a nice picture. The triangle $A_{(0)}$, as all triangles and polygonal sets in Section 1.1.7, will be considered to contain all points inside its boundary. Connecting the midpoints of the sides by straight line segments divides $A_{(0)}$ into four equilateral triangles of side length $\frac{1}{2}$ each. Just as done in the construction of the CANTOR set, we delete the open middle triangle to obtain the set $A_{(1)}$. Continuing in this way with every one of the three equilateral triangles constituting $A_{(1)}$, at the k-th step we arrive at a compact set $A_{(k)}$ consisting of 3^k equilateral triangles of side length $\frac{1}{2^k}$. The total area of these triangles is $\frac{\sqrt{3}}{4}\cdot(\frac{3}{4})^k$. Taking the intersection of all sets $A_{(k)}$, we obtain a compact set $A=\bigcap_{k=1}^{\infty}A_{(k)}$ of two-dimensional LEBESGUE measure $\lim_{k\to\infty}\frac{\sqrt{3}}{4}\cdot(\frac{3}{4})^k=0$ (Figure 1.52). The set A is called the SIERPINSKI *triangle* [Sierpinski, 1916]. It is self-similar and the open triangle $A_{(0)}$ furnishes the set needed for the open set condition. Formula (1.7) gives $\dim_S(A)=\frac{\log 3}{\log 2}\approx 1.585$.

The construction used above may be applied to any initial triangle in place of $A_{(0)}$; a common version of the SIERPINSKI triangle starts out with an isosceles right-angled triangle (Figure 1.53).

Figure 1.52. The approximating set $A_{(6)}$ for the SIERPINSKI triangle.

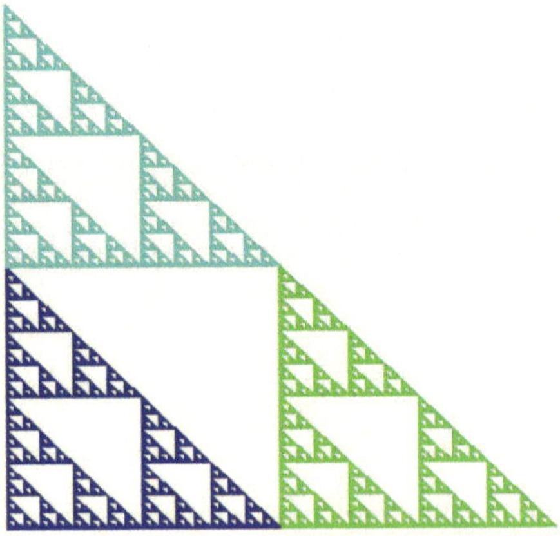

Figure 1.53. The approximating set $A_{(6)}$ for another version of the SIERPINSKI triangle.

1.1.7.2 The LÉVY surface

Recalling how we have come from the CANTOR set to the KOCH curve, we can try to imitate this in two dimensions: replace the deleted triangle in $A_{(1)}$ by three faces of a tetrahedron of side length $\frac{1}{2}$, thereby obtaining a continuous piecewise linear surface $B_{(1)}$ consisting of six equilateral triangles.

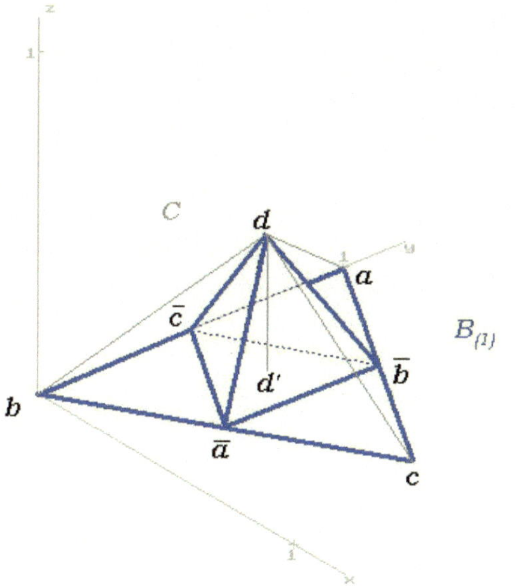

Figure 1.54. The surface $B_{(1)}$ is depicted in blue, the edges *ad*, *bd* and *cd* of the pyramid C and its height $d'd$ in grey.

Now we repeat what we have just done with every one of these equilateral triangles. We obtain a continuous polyhedral surface $B_{(2)}$ consisting of 36 equilateral triangles of side length $\frac{1}{4}$. Our intention is to continue this process in hope of arriving in the limit at a self-similar fractal set B. But what about the open set condition, and can we be sure not to get entangled in some troubling self-intersection of the approximating sets $B_{(k)}$? Fortunately it turns out (and can be checked by elementary geometric considerations as done below) that the open pyramid C, with $A_{(0)}$ as basis and the same top as the first tetrahedron, satisfies the open set condition: its six copies C_j, contracted by a similarity factor $\frac{1}{2}$, built over the three equilateral triangles in $A_{(1)}$ and over the faces of the middle tetrahedron, have pairwise just a face in common and are all contained in C. Continuation of our construction process therefore concerns each time just the part of $B_{(k)}$ inside of the contracted copies of C and leaves their interiors disjoint. By induction we find that the sequence of sets $B_{(k)}$ (which we may also consider as graphs of continuous functions ϕ_k defined on $A_{(0)}$) converges uniformly to a continuous surface B (we shall discuss convergence of sets in this context in Section 2.1), cf. [Lévy, 1938]. The LÉVY *surface* B is also called a *tetrahedral fractal* or KOCH *pyramid*. By formula (1.7) in Section 1.1.3 its dimension is $\dim_S(B) = \frac{\log 6}{\log 2} \approx 2.585$.

A closer study of the construction of B reveals a story with a surprising conclusion. In particular we want to compute the three-dimensional volume of the body $\tilde{B}$ bounded by the fractal surface B and the ground plane (which, in Figures 1.54 and 1.56, coincides with the $[x, y]$-plane).

To this end, let us have a look at the surface $B_{(1)}$. E.g. the face bdc of the pyramid C consists of two halves $bd\bar{a}$ and $cd\bar{a}$. Each of these is an isosceles right-angled triangle. Therefore, so also is their union bdc. Since the three faces adb, bdc and cda are congruent, they consist just of the three half-squares of a cube of side length $\frac{\sqrt{2}}{2}$ joined at the vertex d (Figure 1.55). The height $d'd$ of C is easily computed to be $\frac{1}{\sqrt{6}}$, the area of the basic triangle abc is $\frac{\sqrt{3}}{4}$, and the three-dimensional volume of C is $\frac{1}{12\sqrt{2}}$.

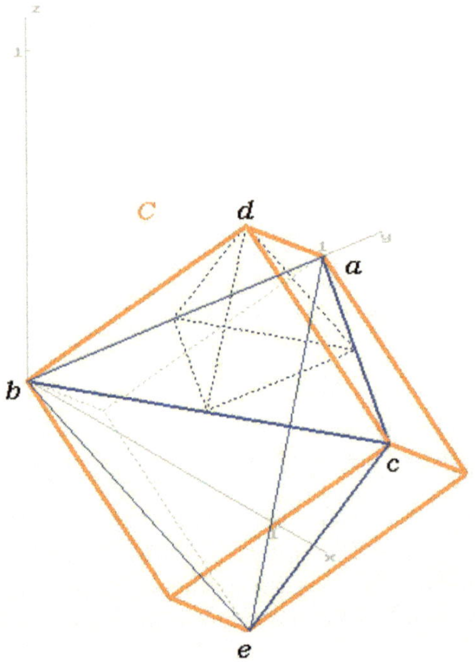

Figure 1.55. The pyramid $C = abcd$ is part of a cube which consists of four pyramids congruent with C, attached to the faces abc, abe, bce and cae of a central tetrahedron $abce$.

Let us now have a look at the surface $B_{(2)}$ resulting from replacing each of the equilateral triangles $a\bar{c}\bar{b}$, $b\bar{a}\bar{c}$, $c\bar{b}\bar{a}$, $\bar{a}\bar{b}d$, $\bar{b}\bar{c}d$, $\bar{c}\bar{a}d$ with a similar copy of $B_{(1)}$, reduced in size by the factor $\frac{1}{2}$. In particular let us have a look at the two copies with basic triangles $b\bar{a}\bar{c}$ and $\bar{c}\bar{a}d$ as in Figure 1.56. The tops d_1 and d_2 of the tetrahedra $a_1b_1c_1d_1$ and $a_2b_2c_2d_2$ coincide with the midpoint of the line segment bd. As a consequence, these tetrahedra intersect just in the edge $b_1d_1 = b_2d_2$, but the corresponding open images of the open pyramid C are disjoint and contained in C, as already remarked above. By induction, $B_{(2)}$ and every following surface $B_{(k)}$ is also contained in (the closed pyramid) C and there is no overlapping of any newly constructed tetrahedra in $B_{(k)}$ apart from common edges.

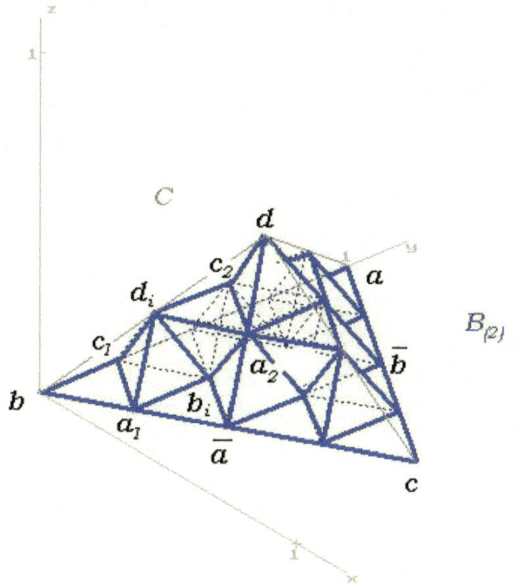

Figure 1.56. The tetrahedra $a_1b_1c_1d_1$ and $a_2b_2c_2d_2$, appearing in the transition from $B_{(1)}$ to $B_{(2)}$, intersect only in the common edge $b_1d_1 = b_2d_2$; both of them are contained in the pyramid $\bar{a}\bar{c}bd$ which is part of the pyramid $C = abcd$ and consists of two similar copies $\bar{a}b\bar{c}d_1$ and $\bar{a}d\bar{c}d_2$ of C, reduced to half of its size.

The content of the first tetrahedron $\bar{a}\bar{b}\bar{c}d$ is $\frac{1}{3} \cdot \frac{\sqrt{3}}{16} \cdot \frac{1}{\sqrt{6}} = \frac{1}{48\sqrt{2}}$. Transition from $B_{(1)}$ to $B_{(2)}$ adds the contents of six tetrahedra half the size. By induction we find that the looked-for volume of $\tilde{B}$ is

$$\frac{1}{48\sqrt{2}} \cdot \left[1 + \frac{3}{4} + \left(\frac{3}{4}\right)^2 + \dots\right] = \frac{1}{48\sqrt{2}} \cdot \frac{1}{1 - \frac{3}{4}} == \frac{1}{12\sqrt{2}},$$

precisely the same as the volume of the pyramid C which entirely contains $\tilde{B}$. As far as three-dimensional LEBESGUE measure is concerned, $\tilde{B}$ entirely fills up the pyramid C, although the surface B is dramatically different from the side faces of C.

We shall not pause here to consider the question whether B is eligible to have any surface measure. Still, this is the case for every approximating surface $B_{(k)}$, consisting of 6^k congruent equilateral triangles of side-length $\frac{1}{2^k}$ and therefore having two-dimensional surface measure $\mathcal{L}^2(B_{(k)}) = 6^k \frac{\sqrt{3}}{2^{2k+2}}$. So $\lim_{k\to\infty} \mathcal{L}^2(B_{(k)}) = \infty$, in analogy to the behavior of the length of the curves approximating e.g. the KOCH island.

There is one more surprise to it: let us proceed analogously as in the case of the KOCH island (Section 1.1.4.1), where we applied the construction of the KOCH curve to every line segment of an equilateral triangle. If now we perform the construction above with each of the four faces of the tetrahedron $abde$ in Figure 1.55 in an outward direction, one might have expected to obtain something like a ball covered by spikes. Instead, the result will be a body, still contained in the original cube but almost filling it up, since it has the same three-dimensional LEBESGUE measure as the cube. The boundary, however, is a closed surface consisting of four copies of the fractal surface B.

1.1.7.3 The CANTOR dust

Take for $A_{(0)}$ the unit square and let s be any positive real number smaller than $\frac{1}{2}$. In order to obtain the first set $A_{(1)}$, delete everything in $A_{(0)}$ which is outside the four small squares of side length s situated at the four corners of $A_{(0)}$. At every step we proceed in the same way with every square so far obtained, reducing the sides of the original square by the factor s. As a result, the set $A_{(k)} \subset A_{(k-1)}$ consists of 4^k squares of side length s^k each and has total area $(2s)^{2k}$. Obviously the limiting set $A = \bigcap_{k=1}^{\infty} A_{(k)}$, called CANTOR *dust*, is self-similar, consisting of four copies similar to itself but reduced with similarity factor s. The interior of $A_{(0)}$ satisfies the open set condition, and formula (1.7) gives $\dim_S(A) = \frac{\log 4}{-\log s}$.

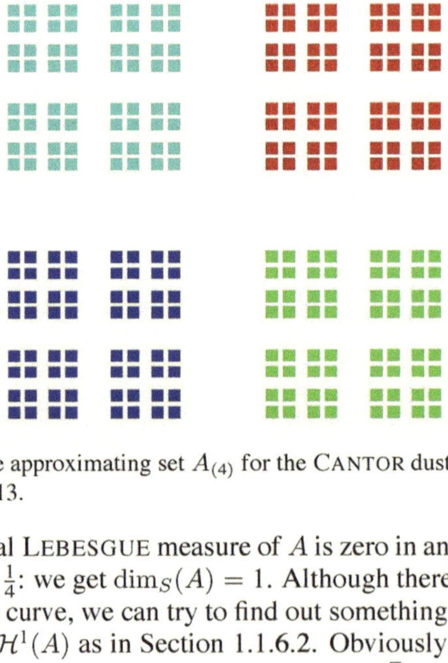

Figure 1.57. The approximating set $A_{(4)}$ for the CANTOR dust A with $s = 0.4$, $\dim_S(A) \approx 1.513$.

The two-dimensional LEBESGUE measure of A is zero in any case, but a special situation happens for $s = \frac{1}{4}$: we get $\dim_S(A) = 1$. Although there is no indication how A may be considered as a curve, we can try to find out something about one-dimensional HAUSDORFF measure $\mathcal{H}^1(A)$ as in Section 1.1.6.2. Obviously A is covered by the 4^k squares constituting $A_{(k)}$, each of which has diameter $\frac{\sqrt{2}}{4^k}$ which becomes arbitrarily small if k is sufficiently large. By (1.9) in Section 1.1.6.2 we obtain $\mathcal{H}^1(A) \leq \sqrt{2}$.

For an estimate from below, consider a somewhat modified CANTOR dust: in $A_{(1)}$ move every small square of side length $\frac{1}{4}$ from its original position in the hook of $A_{(0)}$ a step of length $\frac{1}{4}$ counter-clockwise, say, along the boundary of $A_{(0)}$ to obtain the set $B_{(1)}$. At every step we now proceed in the same way with every square constructed and obtain, as before, the sets $B_{(k)}$ with the same properties as stated above for the sets $A_{(k)}$. In particular, for the set $B = \bigcap_{k=1}^{\infty} B_{(k)}$ we also have $\dim_S(B) = 1$ and $\mathcal{H}^1(B) \leq \sqrt{2}$. Considering again, as in Section 1.1.6.2, the projections p_x and p_y upon the x-axis and y-axis respectively, we see that, for every $k \in \mathbb{N}$, $p_x(B_{(k)})$ and $p_y(B_{(k)})$ are the unit intervals on the x-axis and y-axis respectively. Formula (1.10) in Section 1.1.6.2 implies $\mathcal{H}^1(B) \geq 1$.

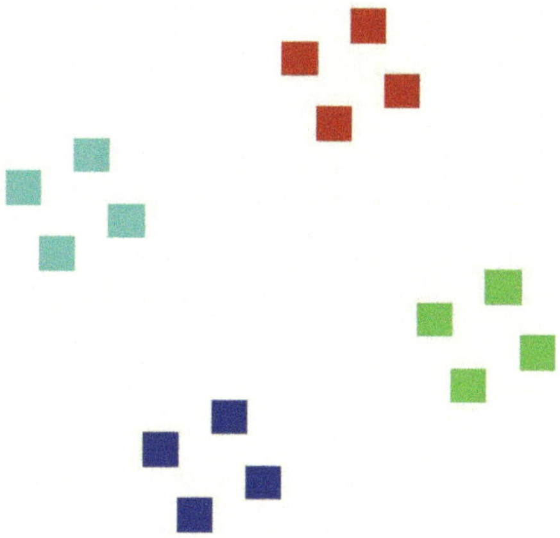

Figure 1.58. The approximating set $B_{(2)}$ for the modified CANTOR dust B with $s = \frac{1}{4}$, $\dim_S(B) = 1$.

There is not much to be seen of CANTOR's dust set B (and A, for that matter), so is there a way to identify, at least analytically, the points of this set? To this end, for $x \in [0, 1]$, denote by x_k $(0 \leq x_k \leq 3)$ the k-th digit in the 4-adic expansion of x, i.e.

$$x = 0.x_1x_2\ldots = \sum_{k=1}^{\infty} \frac{x_k}{4^k}.$$

Let $\mathbb{Q}_4 \subset [0, 1]$ be the set of 4-adic rational numbers

$$x = \sum_{k=1}^{n} \frac{x_k}{4^k} \qquad (n \in \mathbb{N}),$$

and let $\tilde{I} := I \setminus \mathbb{Q}_4$ and $\tilde{Q} := \tilde{I} \times \tilde{I}$. The symbol $\setminus$ denotes set theoretic subtraction, i.e. $\tilde{I}$ is the complement of $\mathbb{Q}_4$ in I. In order to avoid ambiguities, for $x \in \mathbb{Q}_4$ we could e.g. prefer the finite expansion $x = 0.x_1x_2\ldots x_k$ of $x = 0.x_1x_2\ldots x_{k-1}33\ldots$.

The four closed squares B_j $(0 \leq j \leq 3)$ constituting the set $B_{(1)}$ may then be described in the following way:

$$\begin{aligned} B_0 \cap \tilde{Q} &= \{(x,y) \in \tilde{Q} : x_1 = 0,\ y_1 = 2\}, \\ B_1 \cap \tilde{Q} &= \{(x,y) \in \tilde{Q} : x_1 = 1,\ y_1 = 0\}, \\ B_2 \cap \tilde{Q} &= \{(x,y) \in \tilde{Q} : x_1 = 2,\ y_1 = 3\}, \\ B_3 \cap \tilde{Q} &= \{(x,y) \in \tilde{Q} : x_1 = 3,\ y_1 = 1\}. \end{aligned}$$

Correspondingly, for any 4-adic digit x_k let $\tilde{x}_k$ be defined by

$$\tilde{x}_k := \begin{cases} 2 & \text{for } x_k = 0, \\ 0 & \text{for } x_k = 1, \\ 3 & \text{for } x_k = 2, \\ 1 & \text{for } x_k = 3, \end{cases}$$

so that

$$B_j \cap \tilde{Q} = \{(x,y) \in \tilde{Q} : x_1 = j,\ y_1 = \tilde{j}\}.$$

Inductively, for the 4^k closed squares $B_{j_1,\dots,j_k}$ constituting the set $B_{(k)}$, we arrive at the formula

$$B_{j_1,\dots,j_k} \cap \tilde{Q} = \{(x,y) \in \tilde{Q} : x_1 = j_1, \dots, x_k = j_k,\ y_1 = \tilde{j}_1, \dots, y_k = \tilde{j}_k\}.$$

For the fractal $B = \bigcap_{k=1}^{\infty} B_{(k)}$ this furnishes the formula

$$B \cap \tilde{Q} = \{(x,y) : x \in \tilde{I},\ y_k = \tilde{x}_k,\ 1 \leq k < \infty\}.$$

Thus, at least on $\tilde{I}$, the set B turns out to be the graph of a function φ, defined on all but countably many values in $[0,1]$ by $x = \sum_{k=1}^{\infty} \frac{x_k}{4^k} \in \tilde{I} \mapsto \varphi(x) = \sum_{k=1}^{\infty} \frac{\tilde{x}_k}{4^k}$ (it would be possible, if also somewhat cumbersome and not necessary for our purposes, to identify the points in B with abscissas in $\mathbb{Q}_4$). This fact, together with $\dim_S(B) = 1$, nourishes the suspicion that the set B may have something to do with a "short" fractal as met in Section 1.1.6.2. In fact, here is a solution to this puzzle: let the initiator be the unit interval, and let the generator consist of the four intervals of length $\frac{1}{4}$ each in which the set $B_{(1)}$ intersects the boundary of the unit square (this indeed reminds us of the generator used in Section 1.1.6.2). Then the CANTOR dust set B coincides with the fractal produced as done in Section 1.1.6.2.

1.1.7.4 The SIERPINSKI carpet

Again let $A_{(0)}$ be the unit square and let $n \in \mathbb{N}$ be odd and greater than 2. The set $A_{(1)}$ is obtained by removing from $A_{(0)}$ the central square of side length $\frac{1}{n}$. To obtain $A_{(2)}$, in each of the remaining $n^2 - 1$ small squares of side length $\frac{1}{n}$ delete the central square of side length $\frac{1}{n^2}$ and proceed in the same way by induction to construct $A_{(k)}$

for all $k \in \mathbb{N}$. The limiting set $A = \bigcap_{k=1}^{\infty} A_{(k)}$ (especially in the case $n = 3$ often called the SIERPINSKI *carpet*) is self-similar, and the open unit square helps to satisfy the open set condition. The set A has two-dimensional LEBESGUE measure $\mathcal{L}^2(A) = \lim_{k\to\infty} \mathcal{L}^2(A_{(k)}) = \lim_{k\to\infty}(\frac{n^2-1}{n^2})^k = 0$ and dimension $\dim_S(A) = \frac{\log(n^2-1)}{\log n}$.

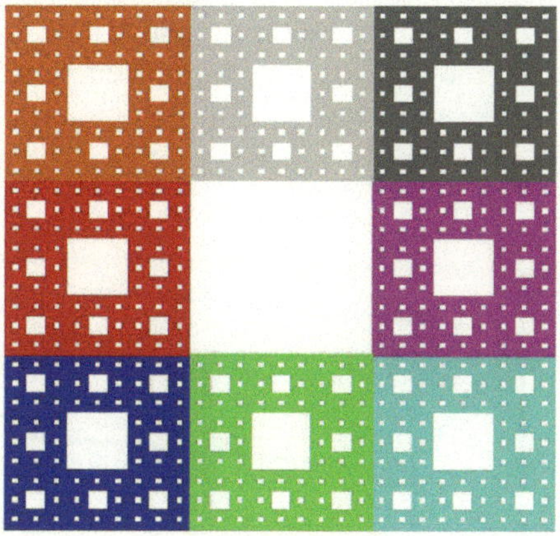

Figure 1.59. The approximating set $A_{(4)}$ for the SIERPINSKI carpet A with $n = 3$ and $\dim_S(A) = \frac{3\log 2}{\log 3} \approx 1.893$.

Especially in the case $n = 3$ a nice modification of what we called the LÉVY surface may be constructed: Replace the deleted small middle square by the 5 faces, above the basis, of a cube of side length $\frac{1}{3}$, to obtain the set $A_{(1)}$ consisting of 13 squares of side length $\frac{1}{3}$. If in each of these small squares the middle square of side length $\frac{1}{3^2}$ is again replaced by 5 protruding faces of cubes of side length $\frac{1}{3^2}$, then eight of these just touch each other pairwise along one edge. Continue this process inductively in the familiar way to obtain the set $A_{(k)}$, a surface consisting of 13^{k-1} copies similar to $A_{(1)}$ but reduced by a similarity factor $(\frac{1}{3})^{k-1}$. The sets $A_{(k)}$ again converge uniformly, in a way which is intuitively clear but will again be discussed in detail in Section 2.1, to a self-similar set A, heuristically only reluctantly considered as a surface of infinite area, with protuberances and caves of arbitrarily small size. The open pyramid C with basis $A_{(0)}$ and top $(0.5, 0.5, 0.5)$ helps to satisfy the open set condition. Formula (1.7) gives $\dim_S(A) = \frac{\log 13}{\log 3} \approx 2.335$.

In contrast to the situation in Section 1.1.7.2 where the body under the LÉVY surface measure-theoretically filled up the pyramid covering it, and admitting the open set condition, this is not the case in the present situation. Since no part of A protrudes above the pyramid C, also no part of A protrudes beyond the similar copies of C reduced by the factor $\frac{1}{3}$ and attached to each of the 13 squares of side length $\frac{1}{3}$ constituting $A_{(1)}$. But there is empty space for instance between C and the union of these 13 smaller copies of C.

1.1.7.5 The MENGER sponge

To keep things short consider the unit cube $A_{(0)} = \{(x_1, x_2, x_3) : 0 \le x_i \le 1,\ 1 \le i \le 3\}$ in $\mathbb{R}^3$. In the direction of each of the coordinate axes, remove a tunnel through $A_{(0)}$, with cross-section the middle square of side length $\frac{1}{3}$ in the face of the cube perpendicular to the coordinate axis. The remains of the unit cube may be divided into 20 smaller cubes of side length $\frac{1}{3}$ each. With every one of these we proceed as before, only reduced by a similarity factor $\frac{1}{3}$, and so on. In the limit we obtain a set $A = \bigcap_{k=1}^{\infty} A_{(k)}$, called the MENGER *sponge* [Menger, 1926], self similar and of dimension $\dim_S(A) = \frac{\log 20}{\log 3} \approx 2.727$, but of three-dimensional LEBESGUE measure $\mathcal{L}^3(A) = \lim_{k\to\infty} \mathcal{L}^3(A_{(k)}) = \lim_{k\to\infty}(\frac{20}{27})^k = 0$.

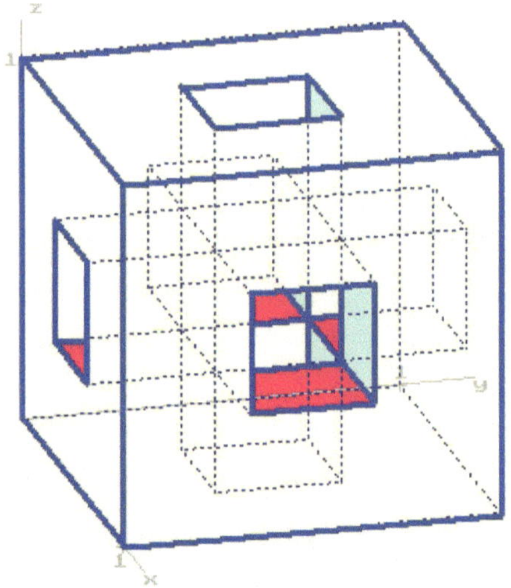

Figure 1.60. The generating set $A_{(1)}$ for the MENGER sponge. Partially visible tunnel faces lying in two planes are marked in colours red and blue respectively.

On each of the faces of the original unit cube there appears a copy of the SIERPINSKI carpet. Transgressing into $\mathbb{R}^4$ one could try again to construct a modified object as in Section 1.1.7.4, but since help from our intuition stops here, we shall stop, too.

1.2 The box-counting dimension

It would certainly seem desirable to be able to attach a dimension not only to self-similar sets in $\mathbb{R}^n$. There are two lines of thought which seem promising: Since in general the set A is no longer the (more or less) disjoint union of a finite number $N_\delta(A)$ of smaller but similar δ-copies of A, we simply count the least number $N_\delta(A)$ of sets of diameter at most δ (> 0) which we need to cover A, and we observe the ratio $\frac{\log N_\delta(A)}{\log \delta}$ as $\delta \to 0$. In an alternative approach, less obvious but in fact more powerful, we attach to each $d \in \mathbb{R}^+ = [0, \infty[$ a measure on $\mathbb{R}^n$ (defined also by means of coverings of sets) and investigate for which value of d this measure seems specially appropriate for the set A.

This section is devoted to a discussion of the first mentioned approach, in which we follow the exposition in [Falconer, 1990]. We start out with the pertinent definitions.

1.2.1 Definition. Let A be a non-empty bounded subset of $\mathbb{R}^n$. Define $N_\delta(A)$ to be the minimal number of subsets of $\mathbb{R}^n$, of diameter not exceeding $\delta(> 0)$, needed to cover A. The *lower* and *upper box-counting dimension* of A are defined respectively by

$$\begin{aligned} \underline{\dim}_B(A) &:= \liminf_{\delta \searrow 0} \frac{\log N_\delta(A)}{-\log \delta}, \\ \overline{\dim}_B(A) &:= \limsup_{\delta \searrow 0} \frac{\log N_\delta(A)}{-\log \delta}. \end{aligned}$$

If both are equal, then

$$\dim_B(A) = \lim_{\delta \searrow 0} \frac{\log N_\delta(A)}{-\log \delta}$$

is called the *box-counting dimension* of A.

Obviously, $\dim_B(A)$ is independent of rotations and translations of the set A. There are equivalent definitions of $\dim_B(A)$ which are handier for the actual computation since they work with restricted classes of sets used in the coverings. Denote by $\mathbb{Z}$ the set of all integers. Let us agree to call a cube of the form $\prod_{i=1}^n [m_i\delta, (m_i + 1)\delta]$ ($m_i \in \mathbb{Z}$, $1 \le i \le n$) a *δ-lattice cube*. In the proofs below frequent use is made of the fact that, for $c > 0$,

$$\lim_{\delta \searrow 0} \frac{\log(c\delta)}{\log \delta} = 1.$$

1.2.2 Theorem. *Let $N'_\delta(A)$ be the minimal number of δ-lattice cubes needed to cover a set $A \subset \mathbb{R}^n$. Then*

$$\begin{aligned} \underline{\dim}_B(A) &= \liminf_{\delta \searrow 0} \frac{\log N'_\delta(A)}{-\log \delta}, \\ \overline{\dim}_B(A) &= \limsup_{\delta \searrow 0} \frac{\log N'_\delta(A)}{-\log \delta}. \end{aligned}$$

Proof. A δ-lattice cube Q has diameter $\delta\sqrt{n}$. Let $\delta < \frac{1}{\sqrt{n}}$. Then $N_{\delta\sqrt{n}}(A) \le N'_\delta(A)$ and therefore

$$\begin{aligned} \underline{\dim}_B(A) &\le \liminf_{\delta \searrow 0} \frac{\log N'_\delta(A)}{-\log \delta}, \\ \overline{\dim}_B(A) &\le \limsup_{\delta \searrow 0} \frac{\log N'_\delta(A)}{-\log \delta}. \end{aligned}$$

On the other hand, every set C of diameter at most δ can be covered by 3^n δ-lattice cubes (take a point in C, cover it by a δ-lattice cube Q and its two neighbours in direction of every coordinate axis; fill up this n-dimensional cross to a cube of side length 3δ: it will cover the whole set C). Therefore $N'_\delta(A) \leq 3^n N_\delta(A)$. This gives us

$$\begin{aligned}\underline{\dim}_B(A) &\geq \liminf_{\delta \searrow 0} \frac{\log N'_\delta(A)}{-\log \delta},\\ \overline{\dim}_B(A) &\geq \limsup_{\delta \searrow 0} \frac{\log N'_\delta(A)}{-\log \delta}.\end{aligned}$$

□

Note that again there is no requirement as to how the coordinate system (and therefore the lattice) should be positioned with respect to the set A. In Theorem 1.2.2 the number $N'_\delta(A)$ may still be replaced e.g. by:

(a) the smallest number of arbitrary cubes of side length δ covering A,

(b) the smallest number of balls with radius δ covering A,

(c) the largest number of disjoint balls with radius δ and center in A.

We omit the proofs since 1) they run along the lines of the proof given for Theorem 1.2.2, and 2) there will be no need for us to use any of the assertions (a), (b), or (c). However, it will turn out to be useful that one may also restrict the family of numbers δ approaching 0:

1.2.3 Theorem. *Suppose the sequence of positive real numbers $\{\delta_k\}_{k=1}^\infty$ converging to zero satisfies, for some $c \in]0,1[$, the inequalities $c \geq \delta_k > \delta_{k+1} \geq c\delta_k$ $(1 \leq k < \infty)$. Then for any set $A \subset \mathbb{R}^n$ one has*

$$\begin{aligned}\underline{\dim}_B(A) &= \liminf_{k \to \infty} \frac{\log N_{\delta_k}(A)}{-\log \delta_k},\\ \overline{\dim}_B(A) &= \limsup_{k \to \infty} \frac{\log N_{\delta_k}(A)}{-\log \delta_k}.\end{aligned}$$

Proof. We shall make use of the facts that $1 > \frac{\delta_{k+1}}{\delta_k} \geq c$ and $0 > \log \frac{\delta_{k+1}}{\delta_k} \geq \log c$. The inequalities

$$\begin{aligned}\underline{\dim}_B(A) &\leq \liminf_{k \to \infty} \frac{\log N_{\delta_k}(A)}{-\log \delta_k},\\ \overline{\dim}_B(A) &\geq \limsup_{k \to \infty} \frac{\log N_{\delta_k}(A)}{-\log \delta_k}\end{aligned}$$

are direct consequences of Definition 1.2.1. For the inequalities in the other directions, given $\delta \in]0, \delta_2[$, choose k such that $\delta_{k+1} \leq \delta < \delta_k$. Then

$$\frac{\log N_\delta(A)}{-\log \delta} \geq \frac{\log N_{\delta_k}(A)}{-\log \delta_{k+1}} = \frac{\log N_{\delta_k}(A)}{-\log \delta_k - \log \frac{\delta_{k+1}}{\delta_k}} \geq \frac{\log N_{\delta_k}(A)}{-\log \delta_k - \log c},$$

$$\underline{\dim}_B(A) = \liminf_{\delta \to 0} \frac{\log N_\delta(A)}{-\log \delta} \geq \liminf_{k \to \infty} \frac{\log N_{\delta_k}(A)}{-\log \delta_k},$$

$$\frac{\log N_\delta(A)}{-\log \delta} \leq \frac{\log N_{\delta_{k+1}}(A)}{-\log \delta_k} = \frac{\log N_{\delta_{k+1}}(A)}{-\log \delta_{k+1} + \log \frac{\delta_{k+1}}{\delta_k}} \leq \frac{\log N_{\delta_{k+1}}(A)}{-\log \delta_{k+1} + \log c},$$

$$\overline{\dim}_B(A) = \limsup_{\delta \to 0} \frac{\log N_\delta(A)}{-\log \delta} \leq \limsup_{k \to \infty} \frac{\log N_{\delta_{k+1}}(A)}{-\log \delta_{k+1}}.$$ □

The hypothesis of Theorem 1.2.3 is satisfied especially if $\delta_k = c^k$ for some $c \in]0, 1[$. This allows us to verify that the self-similarity dimensions of the various fractals which we have got to know so far indeed coincide with their box-counting dimensions:

Let us start with the CANTOR set A. Choose $\delta_k = (\frac{1}{3})^k$. Then $N_{\delta_k}(A) = 2^k$ and $\dim_B(A) = \lim_{k\to\infty} \frac{k \log 2}{k \log 3} = \frac{\log 2}{\log 3}$.

Let now A denote the KOCH curve as in Section 1.1.2. As pointed out in Section 1.1.3 and illustrated in Figure 1.3, the set A may be covered by a sequence of 4^k closed copies of triangles $\{V_{k,i} = a_{k,i}b_{k,i}c_{k,i}\}_{i=1}^{4^k}$ similar to the isosceles triangle $V = abc$, reduced by the similarity factor $\delta_k = 3^{-k}$, and the interiors of the triangles $V_{k,i}$ $(1 \leq i \leq 4^k)$ are disjoint. Since $\operatorname{diam}(V_{k,i}) = \delta_k$ $(1 \leq i \leq 4^k)$ this implies

$$\overline{\dim}_B(A) = \limsup_{k\to\infty} \frac{N_{\delta_k}(A)}{-\delta_k} \leq \frac{\log 4}{\log 3}.$$

Of course other covering sets with diameter less than or equal to 3^{-k} could compete with the sets $V_{k,i}$, but in any case they would have to cover the vertices of each of the triangles $V_{k,i}$. It is readily seen that a closed disc D of radius δ_k with center $a_{k,i}$ meets at most the interiors of $V_{k,i-2}$, $V_{k,i-1}$, $V_{k,i}$ and $V_{k,i+1}$. Any set of diameter less than or equal to δ_k containing $a_{k,i}$ is contained in D. As a consequence we get $N_{\delta_k}(A) \geq \frac{1}{4} \cdot 4^k$ and

$$\frac{\log 4}{\log 3} = \lim_{k\to\infty} \frac{k \log 4 - \log 4}{k \log 3} \leq \liminf_{k\to\infty} \frac{\log N_{\delta_k(A)}}{-\log \delta_k} = \underline{\dim}_B(A).$$

Therefore $\dim_B(A) = \frac{\log 4}{\log 3}$, the same value as the one which has been calculated as $\dim_S(A)$.

What has the open set condition been good for? It has provided for N_{δ_k} to be essentially the number of segments in $A_{(k)}$. If we consider a modified KOCH curve with generator G_4 as in Section 1.1.4.4, then the loops appearing at the vertices admit coverings with sets of diameter less than or equal to δ_k which cover more than one segment of an approximating set $A_{(k)}$ at the same time. Let $N_{S^k}(A)$, as in Section 1.1.3, denote the number of segments in $A_{(k)}$, which is also the number of similar copies of A with similarity factor $\delta_k = \delta(S)^k = (\frac{1}{\sqrt{3}})^k$ contained in A. The just mentioned observation implies $N_{\delta_k}(A) \leq N_{S^k}(A)$. Formula (1.7) is not applicable any more and the dimension of A may well be smaller than $\frac{N_S(A)}{\delta(S)} = \frac{2 \log 2}{\log 3}$.

Let us have a look at the higher-dimensional fractals considered in Section 1.1.7: The dimension of the SIERPINSKI triangle A (Section 1.1.7.1) is most conveniently computed if we consider it with initiating set $A_{(0)}$ not an equilateral triangle but half the unit square to the left below the diagonal (Figure 1.53). Taking $\delta_k = \frac{1}{2^k}$ and relying on Theorem 1.2.2, we obtain $N'_{\delta_k} = 3^k$ and $\dim_B(A) = \lim_{k\to\infty} \frac{k \log 3}{k \log 2} = \frac{\log 3}{\log 2} \approx 1.585$.

For the CANTOR dust A (Section 1.1.7.3) we take $\delta_k = s^k < (\frac{1}{2})^k$. Again an application of Theorem 1.2.2 gives $\dim_B(A) = \lim_{k\to\infty} \frac{k \log 4}{-k \log s} = \frac{\log 4}{-\log s}$. In case of the SIERPINSKI carpet A (Section 1.1.7.4), taking $n > 2$ odd and $\delta_k = \frac{1}{n^k}$, we obtain by the same token $\dim_B(A) = \frac{\log(n^2-1)}{\log n}$. Finally, for the MENGER sponge A (Section 1.1.7.5), taking $\delta_k = \frac{1}{3^k}$, we arrive at $N'_{\delta_k} = 20^k$ and $\dim_B(A) = \frac{\log 20}{\log 3} \approx 2.727$.

We sum up some important properties of the box-counting dimension. (Recall that $\dim_B(A)$ is only defined for bounded subsets $A \subset \mathbb{R}^n$.)

1.2.4 Theorem.

(a) *If $E \subset F$, then $\underline{\dim}_B(E) \leq \underline{\dim}_B(F)$ and $\overline{\dim}_B(E) \leq \overline{\dim}_B(F)$. ("monotony")*

(b) *If Q is a non-empty cube in $\mathbb{R}^n$, then $\dim_B(Q) = n$.*

(c) *If $E \subset \mathbb{R}^n$ is any bounded set, then $\overline{\dim}_B(E) \leq n$.*

(d) *If $E \subset \mathbb{R}^n$ is open, then $\dim_B(E) = n$.*

(e) *$\overline{\dim}_B(E \cup F) = \max(\overline{\dim}_B(E), \overline{\dim}_B(F))$. ("stability")*

(f) *If the function $f : \mathbb{R}^n \to \mathbb{R}^m$ is LIPSCHITZ, i.e. if for some $c > 0$,*

$$|f(x) - f(y)| \leq c|x - y| \qquad \textit{for all } x \in \mathbb{R}^n \textit{ and all } y \in \mathbb{R}^n,$$

and if $E \subset \mathbb{R}^n$, then $\underline{\dim}_B(f(E)) \leq \underline{\dim}_B(E)$ and $\overline{\dim}_B(f(E)) \leq \overline{\dim}_B(E)$.

(g) *If $\overline{E}$ is the closure of $E \subset \mathbb{R}^n$, then $\underline{\dim}_B(\overline{E}) = \underline{\dim}_B(E)$ and $\overline{\dim}_B(\overline{E}) = \overline{\dim}_B(E)$.*

Proof.

(a) The proof is straightforward.

(b) If the side of Q has length s, take $\delta_k = \frac{s}{2^k}$. Then $N'_{\delta_k} = (2^k)^n$ and $\lim_{k\to\infty} \frac{\log N'_{\delta_k}}{-\log \delta_k} = \lim_{k\to\infty} \frac{nk \log 2}{k \log 2 - \log s} = n$.

(c) Take any cube Q containing E and apply (a) and (b).

(d) Take any cube Q contained in E and apply (a) and (b).

(e) The inequality $\overline{\dim}_B(E \cup F) \geq \max(\overline{\dim}_B(E), \overline{\dim}_B(F))$ follows from (a). In order to check the inverse inequality we argue as follows:

$$\begin{aligned} N_\delta(E \cup F) &\leq N_\delta(E) + N_\delta(F) \leq 2\max(N_\delta(E), N_\delta(F)) \\ \frac{\log N_\delta(E \cup F)}{-\log \delta} &\leq \frac{\log 2 + \max(\log N_\delta(E), \log N_\delta(F))}{-\log \delta} \\ \overline{\dim}_B(E \cup F) &\leq \max(\overline{\dim}_B(E), \overline{\dim}_B(F)). \end{aligned}$$

(f) If $E \subset \bigcup_{i=1}^{N_\delta(E)} C_i$ and $\mathrm{diam}(C_i) \le \delta$ for $1 \le i \le N_\delta(E)$, then one has $f(E) \subset \bigcup_{i=1}^{N_\delta(E)} f(C_i)$ and $\mathrm{diam}(f(C_i)) \le c\delta$ $(1 \le i \le N_\delta(E))$. This implies $N_{c\delta}(f(E)) \le N_\delta(E)$ and

$$\frac{\log N_{c\delta}(f(E))}{-\log c\delta} \le \frac{\log N_\delta(E)}{-\log c - \log \delta}.$$

Taking lim inf and lim sup on both sides furnishes the assertion.

(g) If E is covered by N closed sets of diameter at most δ, then the same sets cover $\overline{E}$, i.e. $N_\delta(\overline{E}) = N_\delta(E)$. □

The last assertion (g) implies, e.g. that the countable set of rational numbers in $[0, 1]$ has the same dimension as the interval $[0, 1]$, namely one, while the dimension of one single point is zero. This is disturbing for everybody familiar with measure theory. It is one of the reasons for looking for a concept of dimension which coincides as much as desirable with the box-counting dimension and still satisfies the assertions (a)–(f) of Theorem 1.2.4 but avoids assertion (g).

1.3 The HAUSDORFF dimension

In order to be able to define HAUSDORFF dimension one first has to get acquainted with HAUSDORFF measure $\mathcal{H}^s$, which depends on a non-negative real parameter s, its "dimensionality". For integral values of s, $\mathcal{H}^s$ turns out to coincide with s-dimensional LEBESGUE measure. We have already encountered the one-dimensional HAUSDORFF measure in Section 1.1.6.2. Roughly it measures a subset $A \subset \mathbb{R}^n$ by trying to cover A by countably many sets C_j $(1 \le j < \infty)$ of diameter not exceeding an arbitrarily small $\delta > 0$ in such a way that the sum of the "s-dimensional measures" of "s-dimensional balls" of the same diameter as C_i gets as small as possible. Obviously this imitates the definition of LEBESGUE measure at least of subsets of $\mathbb{R}$, but the terms between the quotes have yet to be defined. There is a formula which does this: the quantity

$$\alpha(s) := \frac{\pi^{s/2}}{\Gamma(\frac{s}{2}+1)} \qquad (s \ge 0)$$

turns out to furnish, for integral s, precisely the s-dimensional LEBESGUE measure of the s-dimensional unit ball (Γ denoting the *gamma function*). Using the well-known formulas $\Gamma(\frac{1}{2}) = \sqrt{\pi}$, $\Gamma(1) = 1$ and $\Gamma(t+1) = t\Gamma(t)$ we get

$$\alpha(0) = 1 \qquad \alpha(1) = 2 \qquad \alpha(2) = \pi \qquad \alpha(3) = \tfrac{4\pi}{3}.$$

It is not quite clear how we should imagine a $\frac{1}{2}$-dimensional unit ball, but according to the formula above its $\frac{1}{2}$-dimensional measure would be $\alpha(\frac{1}{2}) = \frac{\pi^{1/4}}{\Gamma(\frac{5}{4})}$. The rough description above (of $\mathcal{H}^s$) is made precise in the following way.

1.3.1 Definition. Suppose $A \subset \mathbb{R}^n$, $0 \le s < \infty$ and $0 < \delta < \infty$. Then

$$\mathcal{H}^s_\delta(A) := \inf\Big\{\alpha(s) \cdot \sum_{j=1}^{\infty} \Big(\frac{\mathrm{diam}(C_j)}{2}\Big)^s \;:\; A \subset \bigcup_{j=1}^{\infty} C_j,\ \mathrm{diam}(C_j) \le \delta\Big\}.$$

The *s-dimensional* HAUSDORFF *measure* of A is defined by

$$\mathcal{H}^s(A) := \lim_{\delta\to 0} \mathcal{H}^s_\delta(A) = \sup_{\delta>0} \mathcal{H}^s_\delta(A).$$

For convenience, define $\mathcal{H}^s(A) = \infty$ for $s < 0$.

Note again that $\mathcal{H}^s$ is invariant under rotation and translation (this is a special case of the assertion (e) in Theorem 1.3.3 below). Properly speaking, $\mathcal{H}^s$ is an outer measure on $\mathbb{R}^n$, i.e. apart from being non-negative and attributing measure zero to the empty set, for any sequence of subsets $\{A_k\}_{k=1}^\infty \subset \mathbb{R}^n$ it satisfies the inequality

$$\mathcal{H}^s\Big(\bigcup_{k=1}^\infty A_k\Big) \le \sum_{k=1}^\infty \mathcal{H}^s(A_k).$$

As usual, the sets A *measurable with respect to* $\mathcal{H}^s$ are those which satisfy CARATHEODORY's condition

$$\mathcal{H}^s(E) = \mathcal{H}^s(E\cap A) + \mathcal{H}^s(E\setminus A) \qquad \forall\, E\subset \mathbb{R}^n.$$

(The symbol $\forall$ turning up above and in Theorem 1.3.3 denotes the logical "for all".) It turns out that all BOREL sets are $\mathcal{H}^s$-measurable. A proof of the theorem which collects these facts (Theorem 1.3.2) may be found e.g. in [Evans and Gariepy, 1992, p. 61], a text which also inspired the following exposition.

1.3.2 Theorem. *For every $s \ge 0$ and every $n \in \mathbb{N}$, the* HAUSDORFF *measure $\mathcal{H}^s$ is an outer* BOREL *measure on $\mathbb{R}^n$.*

1.3.3 Theorem (Properties of the HAUSDORFF measure).

(a) *$\mathcal{H}^0$ is the counting measure, i.e.* $\mathcal{H}^0(A) = \sum_{p\in A} 1$.

(b) $\mathcal{H}^1(A) = \mathcal{L}^1(A) \quad \forall\, A\subset \mathbb{R}^1$.

(c) $\mathcal{H}^s(A) = 0 \quad \forall\, A\subset\mathbb{R}^n$ *if* $s > n$.

(d) $\mathcal{H}^s(\lambda A) = \lambda^s\cdot\mathcal{H}^s(A)$ *for* $\lambda > 0$.

(e) $\mathcal{H}^s(LA) = \mathcal{H}^s(A)$ *if L is an isometry in $\mathbb{R}^n$, i.e.* $|Lx - Ly| = |x-y|$ *for all* $x\in\mathbb{R}^n$ *and all* $y \in \mathbb{R}^n$.

(f) $\mathcal{H}^n(A) = \mathcal{L}^n(A) \quad \forall\, A\subset\mathbb{R}^n$.

Proof.

(a) Let $p\in\mathbb{R}^n$. Given any $\delta > 0$ by definition we have $\mathcal{H}^0_\delta(\{p\}) = 1$.

(b) Let $A\subset\mathbb{R}^1$ and $\delta > 0$ be given. We shall first show $\mathcal{L}^1(A) \ge \mathcal{H}^1_\delta(A)$ and then $\mathcal{L}^1(A) \le \mathcal{H}^1_\delta(A)$. This will imply $\mathcal{L}^1(A) = \mathcal{H}^1_\delta(A) = \mathcal{H}^1(A)$.

$$\begin{aligned}
\mathcal{L}^1(A) &= \inf\Big\{\sum_{j=1}^\infty(\beta_j-\alpha_j) :\ A\subset\bigcup_{j=1}^\infty\,]\alpha_j,\beta_j[\Big\}\\
&= \inf\Big\{\sum_{j=1}^\infty(\beta_j-\alpha_j) :\ A\subset\bigcup_{j=1}^\infty\,]\alpha_j,\beta_j[,\ 0<\beta_j-\alpha_j\le\delta\Big\}\\
&\ge \mathcal{H}^1_\delta(A).
\end{aligned}$$

For the converse inequality suppose $A \subset \bigcup_{j=1}^{\infty} C_j$, $\bar{\alpha}_j := \inf C_j$, $\bar{\beta}_j := \sup C_j$. We have $\operatorname{diam}(C_j) = \bar{\beta}_j - \bar{\alpha}_j$, and therefore

$$\begin{aligned}\mathcal{H}^1_\delta &= \inf\Big\{\alpha(1)\cdot\sum_{j=1}^{\infty}\frac{\bar{\beta}_j-\bar{\alpha}_j}{2} : A \subset \bigcup_{j=1}^{\infty}[\bar{\alpha}_j,\bar{\beta}_j],\ \bar{\beta}_j-\bar{\alpha}_j \le \delta\Big\}\\ &= \inf\Big\{\sum_{j=1}^{\infty}(\bar{\beta}_j-\bar{\alpha}_j) : A \subset \bigcup_{j=1}^{\infty}[\bar{\alpha}_j,\bar{\beta}_j],\ \bar{\beta}_j-\bar{\alpha}_j \le \delta\Big\}.\end{aligned}$$

Let now also $\varepsilon > 0$ be given and suppose the cover $\{C_j\}_{j=1}^{\infty}$ of A satisfies $\sum_{j=1}^{\infty}(\bar{\beta}_j - \bar{\alpha}_j) < \mathcal{H}^1_\delta(A) + \varepsilon$. Let $\alpha_j := \bar{\alpha}_j - \frac{\varepsilon}{2^{j+1}}$, $\beta_j := \bar{\beta}_j + \frac{\varepsilon}{2^{j+1}}$. Then we have

$$\begin{aligned}A &\subset \bigcup_{j=1}^{\infty}]\alpha_j,\beta_j[,\\ \mathcal{L}^1(A) &\le \sum_{j=1}^{\infty}(\beta_j-\alpha_j) \le \sum_{j=1}^{\infty}(\bar{\beta}_j-\bar{\alpha}_j)+\sum_{j=1}^{\infty}\frac{\varepsilon}{2^j} < \mathcal{H}^1_\delta(A)+2\varepsilon.\end{aligned}$$

(c) We cover the n-dimensional unit cube $Q = [0,1]^n \subset \mathbb{R}^n$ by m^n smaller cubes $Q_{m,k}$ ($1 \le k \le m^n$) with side length $\frac{1}{m}$ ($m \in \mathbb{N}$) and diameter $\frac{\sqrt{n}}{m}$. Given $\delta > 0$ we choose m large enough to satisfy $\frac{\sqrt{n}}{m} < \delta$. For $s > n$ we get

$$\mathcal{H}^s_\delta(Q) \le \alpha(s)\cdot\sum_{k=1}^{m^n}\left(\frac{\sqrt{n}}{2m}\right)^s = \alpha(s)\cdot\frac{\sqrt{n}^{\,s}}{2^s m^{s-n}}.$$

For $m \to \infty$ the last term goes to zero. We conclude

$$\begin{aligned}\mathcal{H}^s_\delta(Q) &= 0 \qquad \text{for all } \delta > 0,\\ \mathcal{H}^s(Q) &= 0.\end{aligned}$$

Since $\mathbb{R}^n$ is a countable union of translated unit cubes this implies $\mathcal{H}^s(\mathbb{R}^n) = 0$.

(d), (e) If $\{C_j\}_{j=1}^{\infty}$ is a covering of A, then $\{\lambda C_j\}_{j=1}^{\infty}$ and $\{LC_j\}_{j=1}^{\infty}$ are coverings of λA and LA respectively.

(f) See e.g. [Evans and Gariepy, 1992, p. 70]. □

As a function of s, the HAUSDORFF measure of a set A behaves somewhat cholerically.

1.3.4 Theorem. *Let $A \subset \mathbb{R}^n$ be given and suppose $0 \le s < t < \infty$. Then:*

(a) $\mathcal{H}^s(A) < \infty \implies \mathcal{H}^t = 0$.

(b) $\mathcal{H}^t(A) > 0 \implies \mathcal{H}^s = \infty$.

Proof.

(a) Let $\delta > 0$ be given and suppose the covering $\{C_j\}_{j=1}^{\infty}$ satisfies $\operatorname{diam}(C_j) \le \delta$ $(1 \le j < \infty)$ and

$$\alpha(s) \cdot \sum_{j=1}^{\infty} \left(\frac{\operatorname{diam}(C_j)}{2}\right)^s \le \mathcal{H}_\delta^s(A) + 1 \le \mathcal{H}^s(A) + 1.$$

We conclude

$$\begin{aligned}
\mathcal{H}_\delta^t(A) &\le \alpha(t) \cdot \sum_{j=1}^{\infty} \left(\frac{\operatorname{diam}(C_j)}{2}\right)^t \\
&= \frac{\alpha(t)}{\alpha(s)} \cdot \alpha(s) \cdot \sum_{j=1}^{\infty} \left(\frac{\operatorname{diam}(C_j)}{2}\right)^s \left(\frac{\operatorname{diam}(C_j)}{2}\right)^{t-s} \\
&\le \frac{\alpha(t)}{\alpha(s)} \cdot (\mathcal{H}^s(A) + 1) \cdot \left(\frac{\delta}{2}\right)^{t-s}.
\end{aligned}$$

Letting $\delta \to 0$ we obtain $\mathcal{H}^t(A) = \lim_{\delta \to 0} \mathcal{H}_\delta^t = 0$.

(b) The assertion (b) is an immediate consequence of (a). □

On second thought, the assertion of Theorem 1.3.4 is not so surprising. We cover the set A with infinitely many sets of small diameter, eventually all smaller than $\delta < 1$, and essentially add up the s-th powers of these diameters. If s gets smaller, then these powers rise in direction of the value 1 which evidently makes the sum diverge. If s gets bigger, then the s-th powers decrease and their sum tends to vanish as $\delta \to 0$. Anyway, the theorem has an interesting consequence: For every set $A \subset \mathbb{R}^n$ there exists an $s_0 \le n$ with the property that

$$\mathcal{H}^s(A) = \begin{cases} \infty & \text{for } s < s_0, \\ 0 & \text{for } s > s_0. \end{cases}$$

This seems to be the "correct" value of s for the set A as far as HAUSDORFF measure is concerned. No wonder we take this for the HAUSDORFF dimension of A.

1.3.5 Definition. The HAUSDORFF *dimension* $\dim_H(A)$ of a set $A \subset \mathbb{R}^n$ is defined by

$$\dim_H(A) := \sup\{s : \mathcal{H}^s(A) = \infty\} = \inf\{s : \mathcal{H}^s(A) = 0\}.$$

Let us compare now HAUSDORFF dimension (defined for all subsets $A \subset \mathbb{R}^n$) with box-counting dimension.

1.3.6 Theorem (Properties of the HAUSDORFF dimension).

(a) $A \subset B \implies \dim_H(A) \le \dim_H(B)$. *("monotony")*

(b) $\dim_H([0,1]^n) = n$.

(c) $A \subset \mathbb{R}^n \implies \dim_H(A) \le n$.

(d) *If* $A \subset \mathbb{R}^n$ *is open, then* $\dim_H(A) = n$.

(e) $\dim_H(\bigcup_{i=1}^\infty A_i) = \sup_{1\le i<\infty} \dim_H(A_i)$. *("countable stability")*

(f) *If the function* $f : \mathbb{R}^n \to \mathbb{R}^m$ *is* LIPSCHITZ (cf. Theorem 1.2.4(f)), *then* $\dim_H(f(A)) \le \dim_H(A)$ *for all* $A \subset \mathbb{R}^n$.

(g) *If* $A \subset \mathbb{R}^n$ *is countable, then* $\dim_H(A) = 0$.

Proof.

(a) $A \subset B$ implies $\mathcal{H}^s(A) \le \mathcal{H}^s(B)$. Therefore we have $\dim_H(A) = \sup\{s : \mathcal{H}^s(A) = \infty\} \le \dim_H(B) = \sup\{s : \mathcal{H}^s(B) = \infty\}$.

(b) $\mathcal{H}^n([0,1]^n) = \mathcal{L}^n([0,1]) = 1$. By Theorem 1.3.3(f) we have $\mathcal{H}^s([0,1])^n = 0$ for $s > n$ and therefore $\dim_H([0,1]^n) = \inf\{s : \mathcal{H}^s([0,1]^s) = 0\} = n$.

(c) For $s > n$ we have $\mathcal{H}^s(\mathbb{R}^n) = 0$. This implies $\dim_H(A) \le n$.

(d) This is a consequence of (a), (b) and (c) in combination with Theorem 1.3.3(f): Every open subset of $\mathbb{R}^n$ contains a coordinate-parallel small cube.

(e) An immediate consequence of (a) is the inequality

$$\dim_H\left(\bigcup_{i=1}^\infty A_i\right) \ge \sup_{1\le i<\infty} \dim_H(A_i).$$

In order to show the reverse inequality it is enough to prove that for every $s > \sup_{1\le i<\infty} \dim_H(A_i)$ we have $\dim_H(\bigcup_{i=1}^\infty A_i) \le s$. Indeed, the supposition implies $\mathcal{H}^s(A_i) = 0$ for all i and therefore also $\mathcal{H}^s(\bigcup_{i=1}^\infty A_i) = 0$.

(f) Suppose

$$|f(x) - f(y)| \le c \cdot |x - y| \qquad \text{for all } (x,y) \in \mathbb{R}^n \times \mathbb{R}^n.$$

We shall show that for all $s \ge 0$ and for all $A \subset \mathbb{R}^n$ this implies

$$\mathcal{H}^s(f(A)) \le c^s \cdot \mathcal{H}^s(A). \tag{1.11}$$

Consequently, for $\mathcal{H}^s(A) = 0$ we also have $\mathcal{H}^s(f(A)) = 0$ and the assertion will follow. Let $\delta > 0$ be given and let $\{C_j\}_{j=1}^\infty$ be a δ-covering of A. Then $\{f(C_j)\}_{j=1}^\infty$ is a $c\delta$-covering of $f(A)$ and

$$\sum_{j=1}^\infty \left(\frac{\operatorname{diam}(f(C_j))}{2}\right)^s \le c^s \cdot \sum_{j=1}^\infty \left(\frac{\operatorname{diam}(C_j)}{2}\right)^s,$$

$$\mathcal{H}^s_{c\delta}(f(A)) \le c^s \cdot \mathcal{H}^s_\delta(A).$$

Letting $\delta \to 0$ we obtain (1.11).

(g) Suppose $A = \bigcup_{i=1}^\infty \{a_i\}$. We have $\mathcal{H}^0(\{a_i\}) = 1$ and therefore $\mathcal{H}^s(\{a_i\}) = 0$ for all $s > 0$ and $i \ge 1$. This implies $\dim_H(\{a_i\}) = 0$ and by (e) $\dim_H(A) = \sup_{1\le i<\infty} \dim_H(\{a_i\}) = 0$. □

We see that HAUSDORFF dimension competes favorably with box-counting dimension and can distinguish between a countable dense set and its closure. So the next theorem will not come as a surprise.

1.3.7 Theorem. *For every bounded subset* $A \subset \mathbb{R}^n$ *one has* $\dim_H(A) \le \underline{\dim}_B(A)$.

Proof. For $\dim_H(A) = 0$ there is nothing to prove. We shall show

$$s < \dim_H(A) \quad \Longrightarrow \quad s \le \underline{\dim}_B(A).$$

The hypothesis implies

$$\lim_{\delta \searrow 0} \mathcal{H}^s_\delta(A) = \mathcal{H}^s(A) = \infty.$$

For sufficiently small $\delta > 0$ we therefore have $\mathcal{H}^s_\delta(A) > 1$. Let such a $\delta < 1$ be given. We can cover A by $N_\delta(A)$ sets of diameter not exceeding δ. Consequently,

$$\begin{aligned}
1 &< \mathcal{H}^s_\delta(A) \le \alpha(s) \cdot N_\delta(A)\Big(\frac{\delta}{2}\Big)^s, \\
0 &< \log \mathcal{H}^s_\delta(A) \le \log \frac{\alpha(s)}{2^s} + \log N_\delta(A) + s \cdot \log \delta, \\
-s \cdot \log \delta &< \log \frac{\alpha(s)}{2^s} + \log N_\delta(A), \\
s &< \frac{\log N_\delta(a) + \log \alpha(s) - s \log 2}{-\log \delta}, \\
s &\le \liminf_{\delta \searrow 0} \frac{\log N_\delta(A)}{-\log \delta} = \underline{\dim}_B(A).
\end{aligned}$$

□

HAUSDORFF and box-counting dimension are peacefully joined in the theorem on which we have already relied so often when trying to get more insight in the structure of self-similar fractals. The proof is rather elaborate, resting on non-trivial parts of measure theory, and is therefore not presented here.

1.3.8 Theorem. *Suppose* S_i $(1 \le i \le m)$ *are similarity mappings with similarity factors* $s_i < 1$. *Suppose the open set condition is satisfied and suppose the bounded non-empty set* $A \subset \mathbb{R}^n$ *satisfies*

$$A = \bigcup_{i=1}^{m} S_i A. \tag{1.12}$$

Then its dimension $d = \dim_H(A) = \dim_B(A)$ *is the solution of the equation*

$$\sum_{i=1}^{m} s_i^d = 1, \tag{1.13}$$

and one has $0 < \mathcal{H}^d(A) < \infty$.

Proof. See e.g. [Falconer, 1990, Section 9.2]. □

Two remarks may help to make this assertion plausible. The first is a heuristic one: the open set condition suggests that the union in (1.12) behaves as a disjoint one, so we expect the equations

$$\mathcal{H}^s(A) = \sum_{i=1}^{m} \mathcal{H}^s(S_i(A)) = \sum_{i=1}^{m} s_i^s \mathcal{H}^s(A)$$

to hold. If we knew that this reasoning was legitimate and that $\mathcal{H}^s(A)$ is finite and positive we could cancel $\mathcal{H}^s(A)$ and would obtain equation (1.13). The trouble is that we do not know this without a considerable amount of mathematical details.

The second remark is more rewarding. If we are willing to accept Theorem 1.3.8, then it helps us to determine the dimension of a fractal even if it is self-similar with similarity factors of different sizes. An example has been discussed in Section 1.1.4.2. There is some trouble to be overcome here, too: equation (1.13) can only be solved numerically, e.g. by NEWTON's method. We look for a zero of the function

$$g(x) = \sum_{i=1}^{n} (s_i)^x - 1$$

with derivative

$$g'(x) = \sum_{i=1}^{n} \log s_i \cdot s_i^x .$$

The looked-for dimension d is then given by

$$x_0 = 0, \qquad x_{k+1} = x_k - \frac{\sum_{i=1}^{n} (s_i)^{x_k} - 1}{\sum_{i=1}^{n} \log s_i \cdot s_i^{x_k}}, \qquad d = \lim_{k\to\infty} x_k .$$

Let us try this for the modified KOCH curve in Section 1.1.4.2 with initiator $A_{(0)} = [0, 1]$ and generator given by the line segments joining consecutively the points

$$p_0 = (0,0) \qquad p_1 = (\tfrac{1}{3}, 0) \qquad p_2 = (\tfrac{1}{3}, \tfrac{1}{5})$$
$$p_3 = (\tfrac{2}{3}, \tfrac{1}{5}) \qquad p_4 = (\tfrac{2}{3}, 0) \qquad p_5 = (0,0).$$

The fractal A satisfies the equation (1.12) with $s_1 = s_3 = s_5 = \frac{1}{3}$ and $s_2 = s_4 = \frac{1}{5}$. NEWTON's method gives as a solution of equation (1.13), which now has the form

$$3\left(\tfrac{1}{3}\right)^d + 2\left(\tfrac{1}{5}\right)^d = 1,$$

the value $d = \dim_H(A) \approx 1.2719$. For the modified KOCH curve studied in Section 1.1.4.2 with a generator consisting of five line segments of equal length $\frac{1}{3}$, we get $(\dim_S(A) =) \dim_H(A) = \frac{\log 5}{\log 3} \approx 1.465$.

2 Iterative function systems

In the first chapter our way of creating fractals relied on iterations of mappings. At the beginning – dealing with initiators and generators – we applied an operator f, replacing every line segment of a given piecewise linear curve by a suitably reduced similar copy of the generator, to an initial set $A_{(0)}$. Repeated application of f produced the sets $A_{(k)}$ which converged point-wise and uniformly to the fractal curve A. We invoked the help of functions ϕ_k on the unit interval of which $A_{(k)}$ were the graphs and which converged uniformly to a function ϕ of which A was the graph. These functions came in handy since they allowed us to pass through the fractal "in time 1" – even if it was space-filling – but for higher dimensional fractals (Section 1.1.7) the handling would have become complicated and we did not pursue this line of thought any further.

This procedure could also be described by the iteration of a finite set of similarities S_j $(1 \leq j \leq n)$, connected with the operator f and arising from its actions on the initiator. Since application of f replaced each segment of the generator by a similar copy of the whole generator, the fractal resulting from infinite repetition of this process became similar to finitely many parts of it, namely the fractals having as initiators the segments of the generator, and these in their turn could be considered as similar copies $S_i A$ of A. In short, we have $A = \bigcup_{j=1}^{n} S_j A$ (suggestive of a snail biting into its own tail). In other words, the fractal A is invariant under application of the operator $F = \bigcup_{j=1}^{n} S_j$. Putting the right side of the above formula for A in place of the last A, and repeating this k times, we get $A = \bigcup_{i=1}^{k} \bigcup_{j_i=1}^{n} S_{j_1} \dots S_{j_k} A$. This formula does not quite hold for the approximating sets $A_{(k)} = \bigcup_{i=1}^{k} \bigcup_{j_i=1}^{n} S_{j_1} \dots S_{j_k} A_{(0)} = F^{(k)} A_{(0)}$, but the similarity of the two formulas nourishes the suspicion that they have something to do with the approximation of A by $A_{(k)}$.

Our aim in this chapter (inspired by the exposition in [Barnsley, 1988]) is (a) to substantiate this suspicion by analyzing the concept of approximation used in this context and (b) to use the corresponding construction to create an even bigger family of fractals by also allowing, in place of S_j, mappings more general than similarities. The first section is devoted to topic (a), the second one to topic (b). In the third section we use the wisdom acquired for creating fractals to our liking in the plane.

2.1 The space of compact subsets of a complete metric space

In practice we shall deal later with $\mathbb{R}^2$ considered as a complete metric space, but we may as well develop our tools in the general setup of an arbitrary complete metric space (X, d). Our intention is to provide the set $\mathcal{K}(X)$ of all compact subsets of X with a metric which makes it again a complete metric space.

2.1.1 Definition. Let $\mathcal{K}(X)$ be defined by $\mathcal{K}(X) := \{A \subset X : A \text{ compact}, \ A \neq \emptyset\}$ (= set of all non-empty compact subsets of X).

2.1.2 Definition. For $a \in X$, $B \in \mathcal{K}(X)$ let the *distance of a from B* be defined by

$$d(a, B) := \min\{d(a, b) : b \in B\}.$$

2.1.3 Definition. For $A \in \mathcal{K}(X)$, $B \in \mathcal{K}(X)$ let the *distance of the set A from the set B* be defined by

$$d(A, B) := \max\{d(a, B) : a \in A\}.$$

Note that in general we have $d(A, B) \neq d(B, A)$. For instance, if $B \subsetneq A$, then $d(A, B) > 0$, $d(B, A) = 0$. So d is not yet suitable for a metric in $\mathcal{K}(X)$.

2.1.4 Definition. For $A \in \mathcal{K}(X)$, $B \in \mathcal{K}(X)$ the HAUSDORFF *distance* of the sets A and B is defined by

$$h(A, B) := \max(d(A, B),\ d(B, A)).$$

2.1.5 Theorem (Properties of the HAUSDORFF distance).

(a) $h(A, A) = 0$.

(b) $A \neq B \iff h(A, B) > 0$.

(c) $h(A, B) = h(B, A)$.

(d) $h(A, C) \leq h(A, B) + h(B, C)$.

Proof. Assertions (a) and (c) are immediate consequences of the definition.

(b) Without loss of generality suppose there is an $a \in A \setminus B$. This implies $d(A, B) \geq d(a, B) > 0$. The converse implication is clear.

(d) It suffices to show $d(A, C) \leq d(A, B) + d(B, C)$ $(\leq h(A, B) + h(B, C))$. This results from the following inequalities, valid for all $a \in A$ and all $b \in B$:

$$\begin{aligned} d(a, C) &= \min\{d(a, c) : c \in C\} \\ &\leq \min\{d(a, b) + d(b, c) : c \in C\} \\ &= d(a, b) + \min\{d(b, c) : c \in C\} \\ &= d(a, b) + d(b, C) \\ &\leq d(a, b) + d(B, C), \\ d(a, C) &\leq d(a, B) + d(B, C). \end{aligned}$$

Taking the maximum over all $a \in A$ on both sides produces the assertion. □

Theorem 2.1.5 guarantees that $(\mathcal{K}(X), h)$ is a metric space. We shall therefore for h also use the term HAUSDORFF *metric*. There is a rather illustrative way to envision the HAUSDORFF distance of two subsets of X: it measures the extent to which they embrace each other (the symbol $\wedge$ denotes the logical "and", the symbol $\exists$ denotes the logical "there exists").

2.1.6 Definition. For $B \in \mathcal{K}(X)$ and $\varepsilon > 0$ let the *ε-hull $B_{\{\varepsilon\}}$ of B* be defined by

$$B_{\{\varepsilon\}} := \{x : d(x, B) \le \varepsilon\} = \{x : \exists b \in B \wedge d(x, b) \le \varepsilon\}.$$

2.1.7 Theorem. $h(A, B) \le \varepsilon \iff A \subset B_{\{\varepsilon\}} \wedge B \subset A_{\{\varepsilon\}}$.

Proof. By definition the inequality $h(A, B) \le \varepsilon$ is equivalent to $d(A, B) \le \varepsilon \wedge d(B, A) \le \varepsilon$. It will therefore suffice to show $d(A, B) \le \varepsilon \iff A \subset B_{\{\varepsilon\}}$. This is a consequence of the equivalence of the following assertions:

$$\begin{aligned} d(A, B) &\le \varepsilon, \\ d(a, B) &\le \varepsilon \quad \forall a \in A, \\ a &\in B_{\{\varepsilon\}} \quad \forall a \in A, \\ A &\subset B_{\{\varepsilon\}}. \end{aligned}$$

□

2.1.8 Lemma. *For $A \in \mathcal{K}(X)$ and $\varepsilon > 0$ the set $A_{\{\varepsilon\}}$ is closed.*

Proof. Let the sequence $\{b_n\}_{n=1}^{\infty} \subset A_{\{\varepsilon\}}$ converge to $b = \lim_{n\to\infty} b_n$. For every $n \in \mathbb{N}$ there exists an $a_n \in A$ satisfying $d(a_n, b_n) \le \varepsilon$. Since A is compact we may without loss of generality suppose that the sequence $\{a_n\}_{n=1}^{\infty}$ converges to some limit $a = \lim_{n\to\infty} a_n$. By the continuity of the distance function d we get $d(a, b) = \lim_{n\to\infty} d(a_n, b_n) \le \varepsilon$. This implies $b \in A_{\{\varepsilon\}}$. □

Now comes the final but somewhat tedious part: we shall convince ourselves that the metric space $(\mathcal{K}(X), h)$ is complete. As is to be expected, we shall have to use the completeness of the space (X, d). Definition 2.1.9 and Theorem 2.1.10 are tools which we shall use. They belong to standard topology. A proof of the theorem is included here in order to increase self-containedness of this presentation.

2.1.9 Definition. A subset $A \subset X$ is called *totally bounded* if given any $\varepsilon > 0$ there exists a finite set $E \subset X$ satisfying $A \subset E_{\{\varepsilon\}}$.

2.1.10 Theorem. *A subset $A \subset X$ is compact iff A is closed and totally bounded.*

Proof. $\Longrightarrow$: Every compact set A is closed. Suppose A is not totally bounded. Then, for some $\varepsilon > 0$, every finite set $\{a_1, \ldots, a_n\} \subset A$ admits an element $a_{n+1} \in A$ satisfying $d(a_i, a_{n+1}) > \varepsilon$ $(1 \le i \le n)$. We may therefore construct inductively a sequence $\{a_i\}_{n=1}^{\infty} \subset A$ satisfying $i < j \implies d(a_i, a_j) \ge \varepsilon$. This sequence has no converging subsequence. This contradicts the compactness of A.

$\Longleftarrow$: Suppose A is closed and totally bounded. Let $\{a_n\}_{n=1}^\infty \subset A$ be a sequence in A. We shall show that this sequence admits a converging subsequence. This will imply compactness of A.

Let $E^{(0)} \subset X$ be a finite subset of X satisfying

$$\{a_n\}_{n=1}^\infty \subset A \subset E^{(0)}_{\{1\}} = \{x \in X : d(x, E^{(0)}) \le 1\}.$$

There has to be a closed ball B_0 with center in $E^{(0)}$ and with radius 1 containing an infinite subsequence $\{a_{0,n}\}_{n=1}^\infty \subset \{a_n\}_{n=1}^\infty$. Any two elements $a_{0,n}$, $a_{0,m}$ in this subsequence have distance $d(a_{0,n}, a_{0,m}) \le 2$. By the same token there is a finite subset $E^{(1)} \subset X$ satisfying

$$\{a_{0,n}\}_{n=1}^\infty \subset A \cap B_0 \subset E^{(1)}_{\{2^{-1}\}} = \{x \in X : d(x, E^{(1)}) \le 2^{-1}\}$$

and a closed ball B_1 with center in $E^{\{1\}}$ and with radius 2^{-1}, containing an infinite subsequence

$$\{a_{1,n}\}_{n=1}^\infty \subset \{a_{0,n}\}_{n=1}^\infty \subset \{a_n\}_{n=1}^\infty.$$

Any two elements $a_{1,n}$, $a_{1,m}$ in this subsequence have distance $d(a_{1,n}, a_{1,m}) \le 1$. In an inductive continuation of this procedure the inductive step starts out with an infinite subsequence $\{a_{k-1,n}\}_{n=1}^\infty \subset \{a_n\}_{n=1}^\infty$ contained in a closed ball B_{k-1} of radius $2^{-(k-1)}$; the inductive step produces a finite set $E^{(k)}$ such that

$$A \cap B_{k-1} \subset E^{(k)}_{\{2^{-k}\}}$$

and a closed ball B_k with center in $E^{(k)}$ and radius 2^{-k} containing an infinite subsequence

$$\{a_{k,n}\}_{n=1}^\infty \subset \{a_{k-1,n}\}_{n=1}^\infty.$$

Any two elements $a_{k,n}$, $a_{k,m}$ in this subsequence have distance $d(a_{k,n}, a_{k,m}) \le 2^{-(k-1)}$.

Now consider the subsequence $\{a_{k,1}\}_{k=0}^\infty \subset \{a_n\}_{n=1}^\infty$. Any two elements $a_{k,1}$, $a_{j,1}$ $(k < j)$ in this subsequence have distance $d(a_{k,1}, a_{j,1}) \le 2^{-(k-1)}$. The subsequence therefore is fundamental and has to converge. □

We shall have to convince ourselves that any fundamental sequence $\{A_n\}_{n=1}^\infty$ in $\mathcal{K}(X)$ admits a limit $A \in \mathcal{K}(X)$. It will turn out that this limit A is the set of all limit points in X of fundamental sequences $\{a_n\}_{n=1}^\infty$ for which $a_n \in A_n$ $\forall n \in \mathbb{N}$. In preparation thereof we shall show that any fundamental sequence $\{a_{n_j}\}_{n=1}^\infty \subset X$ satisfying $a_{n_j} \in A_{n_j}$ for a subsequence of indices $\{n_j\}_{j=1}^\infty \subset \mathbb{N}$ can be replenished to a fundamental sequence $\{a_n\}_{n=1}^\infty$ satisfying $a_n \in A_n$ for all $n \in \mathbb{N}$.

2.1.11 Lemma. *Let $\{A_n\}_{n=1}^\infty$ be a fundamental sequence in $(\mathcal{K}(X), h)$, and let $\{x_{n_j}\}_{j=1}^\infty$ $(n_1 < n_2 < \cdots < n_j < n_{j+1} < \cdots)$ be a fundamental sequence in (X, d) satisfying $x_{n_j} \in A_{n_j}$ $(j \in \mathbb{N})$. Then there is a fundamental sequence $\{\tilde{x}_n\}_{n=1}^\infty$ satisfying*

$$\tilde{x}_n \in A_n \quad \forall n \in \mathbb{N} \qquad \textit{and} \qquad \tilde{x}_{n_j} = x_{n_j} \quad \forall j \in \mathbb{N}.$$

Proof. We shall explicitly construct the looked-for sequence $\{\tilde{x}_n\}_{n=1}^{\infty}$. Let

$$\tilde{x}_{n_j} := x_{n_j} \qquad \forall j \in \mathbb{N}.$$

For $n_{j-1} < n < n_j$ we choose $\tilde{x}_n \in A_n$ in such a way that $d(x_{n_j}, \tilde{x}_n) = d(x_{n_j}, A_n)$. We now have to show: given $\varepsilon > 0$ there is an index N with the property that the inequalities

$$\begin{aligned} N &\leq n_{j-1} < n \leq n_j, \\ N &\leq n_{k-1} < m \leq n_k \end{aligned} \tag{2.1}$$

imply $d(\tilde{x}_n, \tilde{x}_m) < \varepsilon$. To this end choose N such as to satisfy

$$\begin{aligned} d(A_n, A_m) &< \tfrac{\varepsilon}{3} \qquad \forall n \geq N, m \geq N, \\ d(x_{n_j}, x_{n_k}) &< \tfrac{\varepsilon}{3} \qquad \forall n_j \geq N, n_k \geq N. \end{aligned}$$

If n and m satisfy (2.1), then

$$\begin{aligned} d(\tilde{x}_n, \tilde{x}_m) &\leq d(x_{n_j}, \tilde{x}_n) + d(x_{n_j}, x_{n_k}) + d(x_{n_k}, \tilde{x}_m) \\ &\leq d(x_{n_j}, A_n) + \tfrac{\varepsilon}{3} + d(x_{n_k}, A_m) \\ &\leq d(A_{n_j}, A_n) + \tfrac{\varepsilon}{3} + d(A_{n_k}, A_m) \\ &\leq \varepsilon. \end{aligned}$$

□

2.1.12 Theorem. *Let $\{A_n\}_{n=1}^{\infty}$ be a fundamental sequence in $(\mathcal{K}(X), h)$ and define*

$$A := \{x \in X : x = \lim_{n\to\infty} x_n,\ x_n \in A_n\ (n \in \mathbb{N})\}.$$

Then $A = \lim_{n\to\infty} A_n$ with respect to the HAUSDORFF *distance in $\mathcal{K}(X)$. Consequently, the metric space $(\mathcal{K}(X), h)$ is complete.*

Proof. We conduct the proof in five steps.

Claim 1: $A \neq \emptyset$.
Proof: For simplification of the notation let $\varepsilon_k := 2^{-k}$. Choose a monotone sequence of indices $\{N_j\}_{j=1}^{\infty}$ in such a way that

$$h(A_n, A_m) < \varepsilon_j \qquad \forall\, n \geq N_j, m \geq N_j.$$

By Theorem 2.1.7 the conditions $n \geq N_j$, $m \geq N_j$ imply $A_n \subset (A_m)_{\{\varepsilon_j\}}$. Starting with any $x_{N_1} \in A_{N_1}$ we construct a fundamental sequence of elements $x_{N_j} \in A_{N_j}$ as follows: because of $A_{N_1} \subset (A_{N_2})_{\{\varepsilon_1\}}$ we may choose an element $x_{N_2} \in A_{N_2}$ in such a way that $d(x_{N_1}, x_{N_2}) \leq \varepsilon_1$. If $x_{N_j} \in A_{N_j}$ $(1 \leq j \leq k)$ have been chosen in such a way that $d(x_{N_j}, x_{N_{j+1}}) \leq \varepsilon_j$ for $1 \leq j \leq k-1$, then because of $A_{N_k} \subset (A_{N_{k+1}})_{\{\varepsilon_k\}}$ we may again choose $x_{N_{k+1}} \in A_{N_{k+1}}$ in such a way that also $d(x_{N_k}, x_{N_{k+1}}) \leq \varepsilon_k$ is satisfied.

Let now $\varepsilon > 0$ be given. We choose k so large that $\varepsilon_k < \varepsilon$. For $k < i < j$ we get

$$\begin{aligned} d(x_{N_i}, x_{N_j}) &\le d(x_{N_i}, x_{N_{i+1}}) + \cdots + d(x_{N_{j-1}}, x_{N_j}) \\ &\le \sum_{l=i}^{j-1} \varepsilon_l \le \sum_{l=k+1}^{\infty} \varepsilon_l = \varepsilon_k < \varepsilon. \end{aligned}$$

We see that the sequence $\{x_{N_j}\}_{j=1}^{\infty}$ is fundamental in X. By Lemma 2.1.11 we can extend it to a fundamental sequence $\{\tilde{x}_n\}_{n=1}^{\infty}$ which satisfies $\tilde{x}_n \in A_n$ $(n \in \mathbb{N})$ and $\tilde{x}_{N_j} = x_{N_j}$ $(j \in \mathbb{N})$. Its limit point belongs to A which consequently is not empty.

Claim 2: A is closed.
Proof: Suppose the sequence $\{a_m\}_{m=1}^{\infty} \subset A$ converges to the limit $a \in X$. Each element a_m in turn is the limit of a sequence $\{x_{m,n}\}_{n=1}^{\infty}$, $x_{m,n} \in A_n$ $(n \in \mathbb{N})$. Let $\{M_j\}_{j=1}^{\infty}$ be an increasing sequence of indices satisfying

$$d(a, a_m) \le \varepsilon_j \qquad \forall m \ge M_j.$$

Let, furthermore, $\{N_j\}_{j=1}^{\infty}$ be a corresponding increasing sequence of indices satisfying

$$d(a_{M_j}, x_{M_j,N_j}) \le \varepsilon_j \qquad (1 \le j < \infty);$$

then

$$\begin{aligned} d(a, x_{M_j,N_j}) &\le d(a, a_{M_j}) + d(a_{M_j}, x_{M_j,N_j}) \le 2\varepsilon_j, \\ x_{M_j,N_j} &\in A_{N_j}, \\ a &= \lim_{j\to\infty} x_{M_j,N_j}. \end{aligned}$$

By Lemma 2.1.11 we may extend the sequence $\{x_{M_j,N_j}\}_{j=1}^{\infty}$ to a sequence $\{\tilde{x}_n\}_{n=1}^{\infty}$ $(\tilde{x}_n \in A_n,\ n \in \mathbb{N})$ which converges to a. By definition we have $a \in A$. This proves that A is closed.

Claim 3: Given any $\varepsilon > 0$ there is an $N(\varepsilon) \in \mathbb{N}$ such that $n \ge N(\varepsilon) \implies A \subset (A_n)_{\{\varepsilon\}}$.
Proof: Consider any $a \in A$ and a corresponding sequence $\{a_m\}_{m=1}^{\infty} \subset X$ satisfying

$$a = \lim_{m\to\infty} a_m \qquad \text{and} \qquad a_m \in A_m \quad \forall m \in \mathbb{N}.$$

Choose $N(\varepsilon)$ in such a way that $m \ge N(\varepsilon)$, $n \ge N(\varepsilon)$ implies $h(A_m, A_n) < \varepsilon$. Given any $n \ge N(\varepsilon)$, for all $m \ge N(\varepsilon)$ by Theorem 2.1.7 we then have

$$a_m \in A_m \subset (A_n)_{\{\varepsilon\}}.$$

Since the set $(A_n)_{\{\varepsilon\}}$ is closed (Lemma 2.1.8) this implies also $a \in (A_n)_{\{\varepsilon\}}$.

Claim 4: The set A is totally bounded and consequently compact.
Proof: If A were not totally bounded, then, as in the first part of the proof of Theorem 2.1.10, for a suitable $\varepsilon > 0$ we could construct a sequence $\{a_j\}_{j=1}^{\infty} \subset A$ satisfying

$d(a_i, a_j) \geq \varepsilon$ for $i \neq j$. By claim 3 we have $A \subset (A_n)_{\{\varepsilon/3\}}$ for some sufficiently large n. Therefore we could associate with every $a_j \in A$ an element $b_j \in A_n$ such that $d(a_j, b_j) \leq \frac{\varepsilon}{3}$. A_n being compact we may suppose without loss of generality that $\{b_j\}_{j=1}^{\infty}$ converges towards an element $b \in A_n$. For some suitable $N(\varepsilon)$ and all $i \geq N(\varepsilon), j \geq N(\varepsilon)$ we would get

$$\begin{aligned} d(b_i, b_j) &< \frac{\varepsilon}{3}, \\ d(a_i, a_j) &\leq d(a_i, b_i) + d(b_i, b_j) + d(b_j, a_j) < \varepsilon, \end{aligned}$$

a contradiction to our assumption.

Claim 5: $A = \lim_{n\to\infty} A_n$ with respect to HAUSDORFF distance h.
Proof: By Theorem 2.1.7 the claim is equivalent with the following claim. Given any $\varepsilon > 0$ there is an index $N(\varepsilon)$ such that for all $n \geq N(\varepsilon)$

$$A \subset A_{n,\{\varepsilon\}} \qquad \text{and} \qquad A_n \subset A_{\{\varepsilon\}}.$$

Claim 3 has taken care of the first assertion. In order to check the second one let $N(\varepsilon)$ be so large that

$$h(A_m, A_n) \leq \frac{\varepsilon}{2} \qquad \forall m \geq N(\varepsilon), n \geq N(\varepsilon).$$

Let now any $y \in A_n$ $(n \geq N(\varepsilon))$ be given. We are going to show $y \in A_{\{\varepsilon\}}$.

In order to simplify the notation we put now $\varepsilon_k := \frac{\varepsilon}{2^k}$. Since $\{A_n\}_{n=1}^{\infty}$ is a fundamental sequence in $(\mathcal{K}(X), h)$, there is an increasing sequence of indices $\{N_j\}_{j=1}^{\infty}$ $(N_1 \geq n)$ satisfying (due to Theorem 2.1.7)

$$A_k \subset (A_l)_{\{\varepsilon_{j+1}\}} \qquad \forall k \geq N_j, l \geq N_j.$$

The scheme of implications below serves to construct a fundamental sequence $\{x_{N_j}\}_{j=1}^{\infty}$ satisfying $x_{N_j} \in A_{N_j}$ $(1 \leq j < \infty)$:

$$\begin{array}{rcccl} N(\varepsilon) \leq n \leq N_1 & \Rightarrow & A_n \subset (A_{N_1})_{\{\varepsilon_1\}} & \Rightarrow & \exists x_{N_1} \in A_{N_1} : d(y, x_{N_1}) \leq \varepsilon_1, \\ N_1 \leq N_2 & \Rightarrow & A_{N_1} \subset (A_{N_2})_{\{\varepsilon_2\}} & \Rightarrow & \exists x_{N_2} \in A_{N_2} : d(x_{N_1}, x_{N_2}) \leq \varepsilon_2, \\ & & & \vdots & \\ N_{j-1} \leq N_j & \Rightarrow & A_{N_{j-1}} \subset (A_{N_j})_{\{\varepsilon_j\}} & \Rightarrow & \exists x_{N_j} \in A_{N_j} : d(x_{N_{j-1}}, x_{N_j}) \leq \varepsilon_j, \\ & & & \vdots & \end{array}$$

By Lemma 2.1.11 we can extend the fundamental sequence $\{x_{n_j}\}_{j=1}^{\infty}$ to a fundamental sequence $\{\tilde{x}_n\}_{n=1}^{\infty}$ satisfying $\tilde{x}_n \in A_n$ $(n \in \mathbb{N})$ and $\tilde{x}_{N_j} = x_{N_j}$ $(j \in \mathbb{N})$. The limit $x = \lim_{n\to\infty} \tilde{x}_n \in A$ of this sequence satisfies $d(y, x) \leq \sum_{j=1}^{\infty} \varepsilon_j = \varepsilon$ and consequently $y \in A_{\{\varepsilon\}}$. □

2.2 Contractions in a complete metric space

As already indicated at the beginning of this chapter the key to the construction of fractals, generalizing the ones we have got to know so far, is the iterated application of an operator of the form $F = \bigcup_{j=1}^{m} f_j$ to compact sets in $\mathbb{R}^n$, where f_j $(1 \leq j \leq m)$ are suitable mappings $\mathcal{K}(\mathbb{R}^n) \to \mathcal{K}(\mathbb{R}^n)$. Of course one has to impose some condition on the mappings f_j which guarantee that the sequence $\{F^{(k)}(A_{(0)})\}_{k=1}^{\infty}$ converges in the sense of the HAUSDORFF metric, as has seemingly been the case with the similarities S_j employed in the initiator-generator procedure. Again we might as well develop the same convenient means in the general setup of mappings in $\mathcal{K}(X)$ as in Section 2.1. We therefore suppose again that (X, d) is a complete metric space.

2.2.1 Definition. A mapping $f : X \to X$ is called a *contraction* if there is a constant $c \in [0, 1[$ (called a *contraction constant*) such that

$$d(f(x), f(y)) \leq c \cdot d(x, y) \qquad \forall\,(x, y) \in X \times X.$$

Note that, as an immediate consequence of the definition, a contraction is a continuous mapping. In general we do not require the constant c to be optimal. For a similarity in $\mathbb{R}^n$ as in Definition 1.1.3.1, however, a similarity factor $s < 1$ is automatically the smallest contraction constant possible.

2.2.2 Definition. The sequence $\{f^{(k)}(x)\}_{k=0}^{\infty}$ is called the *orbit* of the element $x \in X$ under f. A point $x \in X$ is called a *fixed point* for f if $f(x) = x$.

2.2.3 Theorem. *A contraction $f : X \to X$ has exactly one fixed point and every orbit converges to it.*

Proof. Let c be the contraction constant of f. Let any $x \in X$ and the non-negative integers $m < n$ be given. Then

$$\begin{aligned}
d(f^{(m)}(x), f^{(n)}(x)) &= d(f^{(m)}(x), f^{(m)}f^{(n-m)}(x)) \\
&\leq c^m d(x, f^{(n-m)}(x)), \\
d(x, f^{(n)}(x) &\leq \sum_{k=1}^{n} d(f^{(k-1)}(x), f^{(k)}(x)) \\
&\leq \sum_{k=1}^{n} c^{k-1} \cdot d(x, f(x)) \\
&\leq \frac{1}{1-c} \cdot d(x, f(x)), \\
d(f^{(m)}(x), f^{(n)}(x)) &\leq \frac{c^m}{1-c} \cdot d(x, f(x)). \qquad (2.2)
\end{aligned}$$

The last inequality shows that the sequence $\{f^{(k)}(x)\}_{k=1}^{\infty}$ is fundamental and converges to some limit $x_f = \lim_{k\to\infty} f^{(k)}(x)$. At the same time we get

$$f(x_f) = f(\lim_{k\to\infty} f^{(k)}(x)) = \lim_{k\to\infty} f^{(k+1)}(x) = x_f.$$

For any $y \in X$ satisfying $f(y) = y$ we get $d(x_f, y) = d(f(x_f), f(y)) \leq c \cdot d(x_f, y)$. Because of $0 \leq c < 1$ this implies $y = x_f$. □

2.2.4 Corollary. *Let $f : X \to X$ be a contraction with contraction constant c and let x_f be the fixed point of f. Then for every $x \in X$ one has*

$$d(f^{(m)}(x), x_f) \leq \frac{c^m}{1-c} \cdot d(x, f(x)), \qquad \textit{in particular}$$

$$d(x, x_f) \leq \frac{1}{1-c} \cdot d(x, f(x)).$$

Proof. In formula (2.2) let $n \to \infty$. □

It is well known that a continuous image of a compact set is again compact. For a contraction f we therefore have $f(E) := \{f(x) : x \in E\} \in \mathcal{K}(X)$ for every $E \in \mathcal{K}(X)$. We may therefore consider f also as a map from $\mathcal{K}(X)$ into itself, and we shall do so without extra notation. Fortunately, into its elevated position f also carries along its contraction property.

2.2.5 Theorem. *Let f be a contraction in (X, d) with contraction constant c. Then, as a function on the complete metric space $(\mathcal{K}(X), h)$ into itself, it is again a contraction with contraction constant c.*

Proof. By the definition of the HAUSDORFF metric h it suffices to check the following for $A \in \mathcal{K}(X)$, $B \in \mathcal{K}(X)$:

$$\begin{aligned} d(f(A), f(B)) &= \max\{d(f(a), f(B) : a \in A\} \\ &= \max\{\min\{d(f(a), d(f(b) : b \in B\} : a \in A\} \\ &\leq \max\{\min\{cd(a, b) : b \in B\} : a \in A\} \\ &= c \cdot d(A, B). \end{aligned}$$ □

If we only apply the iterates of a single contraction f to a set $A \in \mathcal{K}(X)$, then the images simply contract into the fixed point of f. In order to get more than that we therefore have to invoke the help of some more contractions f_j $(1 \leq j \leq J)$ and combine the single image sets $f_j(A)$ to the bigger set $\bigcup_{j=1}^{J} f_j(A)$, also called *collage* (of the set A under the contractions f_j $(1 \leq j \leq J)$; the term may also be used for the mapping $F = \bigcup_{j=1}^{J} f_j)$. This procedure turns out to be very effective and has therefore given rise to an extra definition (2.2.6), but in order to see this we have to provide some intermediate results.

2.2.6 Definition. A finite set of contractions f_j $(1 \leq j \leq J)$ in a complete metric space (X, d) is called an *iterative function system (IFS)*; the functions f_j are called *collage mappings*.

2.2.7 Lemma. *For compact subsets A, B and C of X one has*

$$d(A \cup B, C) = \max\big\{d(A, C),\ d(B, C)\big\}.$$

Proof.

$$\begin{aligned} d(A\cup B,C) &= \max\{d(x,C) : x\in A\cup B\} \\ &= \max\{\max\{d(x,C) : x\in A\}, \max\{d(x,C) : x\in B\}\} \\ &= \max\{d(A,C), d(B,C)\}. \end{aligned}$$

□

2.2.8 Lemma. *For compact subsets A, B, C and D of X one has*

$$h(A\cup B, C\cup D) \le \max\{h(A,C), h(B,D)\}.$$

Proof.

$$\begin{aligned} &h(A\cup B, C\cup D) \\ &= \max\{d(A\cup B, C\cup D), d(C\cup D, A\cup B)\} \\ &= \max\{\max\{d(A, C\cup D), d(B, C\cup D)\}, \max\{d(C, A\cup B), d(D, A\cup B)\}\} \\ &\quad \text{(by Lemma 2.2.7)} \\ &\le \max\{\max\{d(A,C), d(B,D)\}, \max\{d(C,A), d(D,B)\}\} \\ &= \max\{h(A,C), h(B,D)\}. \end{aligned}$$

□

There is no need for confusion as to whether in Lemma 2.2.8 one has to combine A with C or D and B with D or C; simply take your choice. We can now harvest the fruits of our endeavors so far.

2.2.9 Theorem. *Let $\{f_j\}_{j=1}^J$ be an IFS in (X,d) and let the map $F : \mathcal{K}(X) \to \mathcal{K}(X)$ be defined by*

$$F(A) := \bigcup_{j=1}^{J} f_j(A) \qquad (A \in \mathcal{K}(X)). \tag{2.3}$$

Then F is a contraction in $(\mathcal{K}(X), h)$ with contraction constant $c = \max\{c_j : 1 \le j \le J\}$, where c_j is a contraction constant of f_j for $1 \le j \le J$.

Proof. It suffices to prove the assertion for $J = 2$ and to proceed by induction. We find

$$\begin{aligned} h(F(A), F(B)) &= h(f_1(A)\cup f_2(A), f_1(B)\cup f_2(B)) \\ &\le \max\{h(f_1(A), f_1(B)), h(f_2(A), f_2(B))\} \quad \text{(by Lemma 2.2.8)} \\ &\le \max\{c_1\cdot h(A,B), c_2\cdot h(A,B)\} \\ &\le c\cdot h(A,B). \end{aligned}$$

□

2.2.10 Theorem. *Let $\{f_j\}_{j=1}^J$ be an IFS in (X,d) and let $F : \mathcal{K}(X) \to \mathcal{K}(X)$ be given by (2.3). Then there is a unique fixed point $A \in \mathcal{K}(X)$ of F, called the "attractor of F", and with respect to the* HAUSDORFF *metric h in X one has*

$$\begin{aligned} A &= \lim_{k\to\infty} F^{(k)}(B) \qquad \textit{for all } B \in \mathcal{K}(X) \\ &= F(A) \\ &= \bigcup_{j=1}^J f_j(A). \end{aligned}$$

Proof. Apply Theorem 2.2.3 to the contraction F in the complete metric space $(\mathcal{K}(X), h)$. □

2.2.11 Theorem ("Collage Theorem"). *Let $\{f_j\}_{j=1}^J$ be an IFS in (X,d) and let $F : \mathcal{K}(X) \to \mathcal{K}(X)$ be given by (2.3) and let $c = \max\{c_j : 1 \le j \le J\}$, where c_j is a contraction constant of f_j for $1 \le j \le J$. Let A be the attractor of F and let $B \in \mathcal{K}(X)$ be given. Then*

$$h(B,A) \le \frac{1}{1-c} \cdot h(B, F(B)).$$

Proof. Apply Corollary 2.2.4 to the contraction F in the complete metric space $(\mathcal{K}(X), h)$. □

It will be convenient to have one more piece of knowledge furnishing insight in the continuous dependence of an attractor on changes in the IFS producing it.

2.2.12 Theorem. *Let (T, d_T) (the parameter space) be a metric space and let $\{\{f_{t,j}\}_{j=1}^J : t \in T\}$ be a family of IFS in (X,d) with contraction constants $c_t \le c < 1$ and attractors A_t respectively. Suppose for some constant C and for every $x \in X$ one has*

$$d(f_{t,j}(x), f_{s,j}(x)) \le C \cdot d_T(t,s) \qquad (t \in T,\ s \in T,\ 1 \le j \le J).$$

Then

$$h(A_t, A_s) \le \frac{C}{1-c} \cdot d_T(t,s).$$

Proof. For $t \in T$ let F_t be defined as in (2.3). By Corollary 2.2.4 we have

$$h(A_t, A_s) \le \frac{1}{1-c} \cdot h(A_t, F_s(A_t)) = \frac{1}{1-c} \cdot h(F_t(A_t), F_s(A_t)).$$

It suffices to show that for every $B \in \mathcal{K}(X)$ we have

$$h(F_t(B), F_s(B)) \le C \cdot d_T(t,s).$$

By Lemma 2.2.8 we know

$$h\Big(\bigcup_{j=1}^J f_{t,j}(B), \bigcup_{j=1}^J f_{s,j}(B)\Big) \le \max\big\{h(f_{t,j}(B), f_{s,j}(B)) : 1 \le j \le J\big\}.$$

Therefore it suffices to show

$$d(f_{t,j}(B), f_{s,j}(B)) \leq C \cdot d_T(t,s).$$

To this end we argue as follows: given $x \in f_{t,j}(B)$ let $x_{t,j} \in B$ be chosen such that $f_{t,j}(x_{t,j}) = x$. Then

$$\begin{aligned} d(x, f_{s,j}(x_{t,j})) &= d(f_{t,j}(x_{t,j}), f_{s,j}(x_{t,j})) \leq C \cdot d_T(t,s), \\ d(x, f_{s,j}(B)) &\leq C \cdot d_T(t,s), \\ d(f_{t,j}(B), f_{s,j}(B)) &\leq C \cdot d_T(t,s). \end{aligned}$$

□

In the examples which we shall meet (Sections 2.3.4.5, 2.3.4.6 and 2.3.7) the parameter space is some bounded interval $T \subset \mathbb{R}$ and X is some bounded subset of $\mathbb{R}^2$, as e.g. the unit square, while $d(x,y) = |x-y|$ and $d_T(t,s) = |t-s|$. The coordinate functions $g_{j,i}$ $(i = 1,2)$ of the mappings $f_{t,j} : x \mapsto f_{t,j}(x) = (g_{j,1}(t,x), g_{j,2}(t,x))$ have the property that the partial derivatives $\frac{\partial g_{j,i}}{\partial t}$ are continuous and therefore uniformly bounded functions on the compact space $T \times X$. By the mean value theorem one has

$$\frac{g_{j,i}(t,x) - g_{j,i}(s,x)}{t-s} = \frac{\partial g_{j,i}}{\partial t}(t', x) \qquad (t' \in]s,t[\subset T).$$

It is readily seen that this guarantees that the hypotheses of Theorem 2.2.12 are satisfied.

2.3 Affine iterative function systems in $\mathbb{R}^2$

As MANDELBROT has pointed out, fractals are found abound and seem to be the result of a favorite construction principle in nature. This is reflected in the two-dimensional images with which to a large extent we perceive the world around us. Striking examples are provided by the BARNSLEY fern and other fractals hardly distinguishable from images of objects in nature.

Our intention in this section is to experience the interdependence between IFS's and the fractals produced by them, and then – following the visible traces laid out by nature and with the help of mathematics – to construct fractals which look approximately like a given compact set $B \subset \mathbb{R}^2$. As a tool we use suitable IFS's, preferentially made up by contractions which mathematically are easy to handle, as were the similarities in Chapter 1. A much wider and in fact sufficiently wide family of contractions are those provided by affine maps in $\mathbb{R}^2$.

2.3.1 Definition. An *affine map* f in $\mathbb{R}^2$ is given by

$$f(x) := Lx + b \qquad (x \in \mathbb{R}^2), \tag{2.4}$$

where L is a linear map, given by a matrix also denoted by L, and $b \in \mathbb{R}^2$.

When computations require a more detailed notation we shall write vectors as columns and use the well-known matrix operations transposition (indicated by ^{t}), multiplication and addition. E.g. formula (2.4) is then written in the form

$$\begin{pmatrix} f(x)_1 \\ f(x)_2 \end{pmatrix} = \begin{pmatrix} l_{1,1} & l_{1,2} \\ l_{2,1} & l_{2,2} \end{pmatrix} \begin{pmatrix} x_1 \\ x_2 \end{pmatrix} + \begin{pmatrix} b_1 \\ b_2 \end{pmatrix}.$$

A *rotation* about an angle φ is given by an orthogonal matrix

$$O = \begin{pmatrix} \cos\varphi & -\sin\varphi \\ \sin\varphi & \cos\varphi \end{pmatrix}.$$

If in (2.4) we have $L = s \cdot O$, then f is a similarity with similarity factor s.

In order to decide whether an affine map is a contraction, we have to compute a contraction constant. Here we are even able to get the smallest and thereby optimal one.

2.3.2 Theorem. *The smallest contraction constant of the affine map f as in* (2.4) *is the square root of the largest eigenvalue of the symmetric matrix L^tL.*

Proof. As is well known in linear algebra, there exists an orthogonal matrix O diagonalizing the symmetric matrix L^tL, i.e. satisfying

$$O^t(L^tL)O = \begin{pmatrix} c_1 & 0 \\ 0 & c_2 \end{pmatrix} = D.$$

The coefficients c_1 and c_2 are the (necessarily real and non-negative) eigenvalues of L^tL. Without loss of generality we assume $c_1 \geq c_2$. The distance in $\mathbb{R}^2$ of two vectors $x \in \mathbb{R}^2$ and $y \in \mathbb{R}^2$ is given by $d(x,y) = |x-y| = \sqrt{(x_1-y_1)^2+(x_2-y_2)^2} = \sqrt{(x-y)^t\cdot(x-y)}$. In order to simplify the notation we shall write $O^t(x-y) = u$. Note that $|u|^2 = |x-y|^2$. We get

$$\begin{aligned} |f(x)-f(y)|^2 &= (L(x-y))^t(L(x-y)) &&= (x-y)^tL^tL(x-y) \\ &= (x-y)^tODO^t(x-y) &&= u^tDu \\ &= c_1u_1^2+c_2u_2^2 \leq c_1|u|^2 &&= c_1|x-y|^2. \end{aligned}$$

Finally note that for $x-y = O(1,0)^t$ we have equality in place of $\leq$, so $\sqrt{c_1}$ is indeed the optimal contraction constant. □

In order that f be a contraction in the sense of Definition 2.3.1 this value c_1 must be smaller than 1. In this case the matrix L cannot have the eigenvalue 1. The next assertion is then an immediate consequence of formula (2.4).

2.3.3 Theorem. *If the affine map f as in* (2.4) *is a contraction, then its fixed point is $x_f = (I-L)^{-1}b$.*

As a consequence, formula (2.4) may be reformulated incorporating the fixed point to

$$f(x) = Lx + (I - L)x_f.$$

Before making use of the broadened possibilities to construct fractals in $\mathbb{R}^2$ let us see whether there is any reward in applying our newly gained knowledge to the construction of the fractals we are already familiar with. In fact we now obtain a comfortable quantification of the degree of approximation of the resulting fractals.

Consider first the version of the SIERPINSKI triangle A discussed in Section 1.1.7.1 and Theorem 1.2.3 (Figure 1.53). Knowing its self-similarity properties we see that it is mapped onto itself by the operator $F = \bigcup_{j=1}^{3} f_j$ where the affine mappings f_j are given as in (2.4) by (I denoting the unit matrix)

$$L_1 = L_2 = L_3 = \tfrac{1}{2}I = \begin{pmatrix} 0.5 & 0 \\ 0 & 0.5 \end{pmatrix},$$

$$b_1 = \begin{pmatrix} 0 \\ 0 \end{pmatrix}, \quad b_2 = \begin{pmatrix} 0.5 \\ 0 \end{pmatrix}, \quad b_3 = \begin{pmatrix} 0 \\ 0.5 \end{pmatrix}.$$

All mappings f_j are similarities with similarity factor $s_j = \frac{1}{2}$ which therefore is also a contraction constant for $F = \bigcup_{j=1}^{3} f_j$. Applying the iterations of F e.g. to the unit square B as initial set, we may estimate the approximation of A by $F^{(m)}(B)$ by means of Corollary 2.2.4. The set $F(B)$ consists of three squares with side length $\frac{1}{2}$ each. For $h(F^{(m)}(B), A)$ we get

$$\begin{aligned} h(B, F(B)) &= \max\{d(B, F(B)),\, d(F(B), B)\} = d(B, F(B)) \\ &= \max\{d(b, F(B)) : b \in B\} = \frac{1}{2}, \\ h(F^{(m)}(B), A) &\le \frac{s^m}{1-s}\, h(B, F(B)) = \frac{1}{2^m}. \end{aligned}$$

The original version of the SIERPINSKI triangle as in Section 1.1.7.1 is obtained by changing b_3 to $(\frac{1}{4}, \frac{\sqrt{3}}{4})^t$. In the last approximation estimate, only $h(B, F(B))$ changes from $\frac{1}{2}$ to the length of a diagonal of a rectangle with sides $\frac{1}{4}$ and $\frac{1}{2} - \frac{\sqrt{3}}{4}$, approximately equal to 0.26.

The collage mappings f_j for the CANTOR dust (Section 1.1.7.2) and the SIERPINSKI carpet (Section 1.1.7.3) are as easily identified. They all are similarities with equal similarity factors which leave the directions of the coordinate axes unchanged.

The KOCH curve A is obtained by means of the mapping $F = \bigcup_{j=}^{4} f_j$ where the f_j

are similarities with common similarity factor $s_j = 1/3$ given by

$$L_1 = L_4 = \tfrac{1}{3}I = \begin{pmatrix} \frac{1}{3} & 0 \\ 0 & \frac{1}{3} \end{pmatrix}, \qquad b_1 = \begin{pmatrix} 0 \\ 0 \end{pmatrix}, \quad b_4 = \begin{pmatrix} \frac{2}{3} \\ 0 \end{pmatrix},$$

$$L_2 = \begin{pmatrix} \frac{1}{6} & -\frac{\sqrt{3}}{6} \\ \frac{\sqrt{3}}{6} & \frac{1}{6} \end{pmatrix}, \qquad b_2 = \begin{pmatrix} \frac{1}{3} \\ 0 \end{pmatrix},$$

$$L_3 = \begin{pmatrix} \frac{1}{6} & \frac{\sqrt{3}}{6} \\ -\frac{\sqrt{3}}{6} & \frac{1}{6} \end{pmatrix}, \qquad b_3 = \begin{pmatrix} \frac{1}{2} \\ \frac{\sqrt{3}}{6} \end{pmatrix}.$$

If we choose as initial set B the unit interval on the x-axis, then $d(B, F(B)) = \frac{1}{4\sqrt{3}}$, $h(B, F(B)) = d(F(B), B) = \frac{1}{2\sqrt{3}}$. By Corollary 2.2.4 we get $h(F^{(m)}(B), A) \leq \frac{1}{3^m(1-1/3)} h(B, F(B)) = \frac{1}{4 \cdot 3^{m-1}\sqrt{3}}$. For this concept of approximation we do not need the help of a function ϕ as in Section 1.1.2. On the other hand, the IFS method does not give any information on how to consider A as a curve.

In case of the short fractal set A in Section 1.1.6.2 with HAUSDORFF dimension 1, it comes in handy that we do not need to consider it as the graph of a function (which would have to be highly discontinuous) in order to have it properly defined as a limit set. It is the product of the IFS consisting of two similarities $\{f_1, f_2\}$ with common similarity factor $\frac{1}{2}$, given as in (2.4) by

$$L_1 = \begin{pmatrix} \frac{1}{2} & 0 \\ 0 & \frac{1}{2} \end{pmatrix}, \qquad b_1 = \begin{pmatrix} 0 \\ 0 \end{pmatrix},$$

$$L_2 = \begin{pmatrix} 0 & -\frac{1}{2} \\ \frac{1}{2} & 0 \end{pmatrix}, \qquad b_2 = \begin{pmatrix} 1 \\ \frac{1}{2} \end{pmatrix}.$$

Taking again as initial set B the unit interval on the x-axis we obtain $d(B, F(B) = \frac{1}{2}$, $h(B, F(B)) = d(F(B), B) = 1$ and $h(F^{(m)}(B), A) \leq \frac{1}{2^{m-1}}$.

2.3.4 Some examples of fractals constructed by means of IFS

Some information concerning the illustrations in the present Chapter 2, produced by means of a Turbo Pascal program, seems to be in order. For every screen pixel (i, k) carrying a colour value, indicating a point $z(i, k)$ of the set to which the IFS mapping $F = \bigcup_{j=1}^{n} f_j$ is applied, this point $z(i, k)$ is mapped by every one of the mappings f_j $(1 \leq j \leq n)$ into its image $f_j(z(i, k))$. The corresponding pixels (i_j, k_j) are stored consecutively in a file together with the index j, indicating at the same time the colour corresponding to the mapping f_j. These colours are blue (1), green (2), cyan (3), red (4), magenta (5), brown (6), light gray (7), dark gray (8), and light blue (9). Finally the file entries are recalled consecutively to produce the new screen content. This implies that the images of the original set under the mappings f_j with higher index j in general cover those for smaller indices.

In the figures we shall adopt the terminology "initial set" for the set $A_{(0)}$, the orbit of which under the IFS mapping $F = \bigcup_{j=1}^{n} f_j$ converges to the attracting fractal A, and "generating set" for the set $A_{(1)} = F(A_{(0)})$ which characterizes the mapping F. This corresponds with the terminology "initiator" and "generator" in Chapter 1.

2.3.4.1 A cross

Consider the five similarities f_j reducing the unit square by the factor $\frac{1}{3}$ and placing the images in form of a cross. As in (2.4) these affine mappings are given by matrices L_j and translation vectors b_j $(1 \leq j \leq 5)$ as follows:

$$L_j = \begin{pmatrix} \frac{1}{3} & 0 \\ 0 & \frac{1}{3} \end{pmatrix} \quad (1 \leq j \leq 5),$$

$$b_1 = \begin{pmatrix} \frac{1}{3} \\ 0 \end{pmatrix}, \quad b_2 = \begin{pmatrix} 0 \\ \frac{1}{3} \end{pmatrix}, \quad b_3 = \begin{pmatrix} \frac{1}{3} \\ \frac{1}{3} \end{pmatrix}, \quad b_4 = \begin{pmatrix} \frac{2}{3} \\ \frac{1}{3} \end{pmatrix}, \quad b_5 = \begin{pmatrix} \frac{1}{3} \\ \frac{2}{3} \end{pmatrix}.$$

If we choose the boundary of the unit square as the initial set $A_{(0)}$, then $A_{(1)} = F(A_{(0)})$ looks as in Figure 2.1, $A_{(2)} = F^{(2)}(A_{(0)})$ looks as in Figure 2.2, and $A_{(6)} = F^{(6)}(A_{(0)})$ looks as in Figure 2.3, already rather close to the expected attractor. But Theorem 2.2.10 has stated that any other initial set $A_{(0)}$ would do as well. So Figures 2.4 and 2.5 illustrate the sets $A_{(1)}$ and $A_{(2)}$ if $A_{(0)}$ is a full circle, while the next Figures 2.6–2.9 illustrate these sets if $A_{(0)}$ is a quarter circle or a triangle respectively. In every case, the set $A_{(6)}$ already looks as in Figure 2.3.

The present IFS satisfies the open set condition (just consider the images of the open unit square). Therefore Theorem 1.3.8 applies and the dimension of the resulting fractal is $\dim_H(A) = \dim_B(A) = \frac{\log 5}{\log 3} \approx 1.465$.

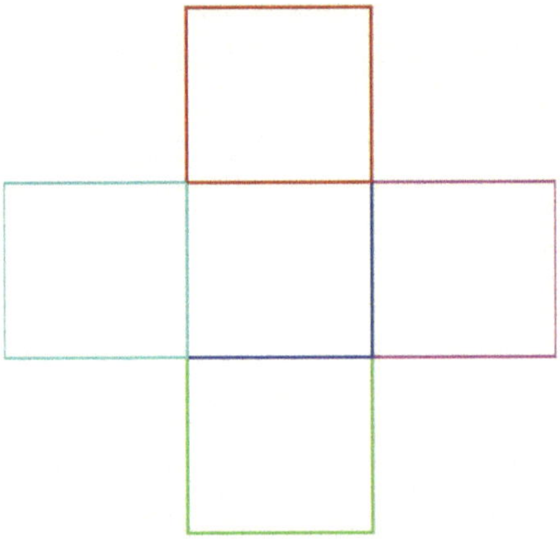

Figure 2.1. The generating set $A_{(1)}$ for the cross, consisting of five images of the initial set $A_{(0)}$ which consists of the four sides of the unit square.

Figure 2.2. The approximating set $A_{(2)}$ for the cross, with initial set as above.

Figure 2.3. The approximating set $A_{(6)}$ for the cross.

Figure 2.4. The generating set $A_{(1)}$ for the cross, if the initial set $A_{(0)}$ is a circular disc.

Figure 2.5. The approximating set $A_{(2)}$ for the cross, with initial set a circular disc.

Figure 2.6. The generating set $A_{(1)}$ for the cross, if the initial set $A_{(0)}$ is a quarter circle.

Figure 2.7. The approximating set $A_{(2)}$ for the cross, with initial set as above.

Figure 2.8. The generating set $A_{(1)}$ for the cross, if the initial set $A_{(0)}$ is a triangle.

Figure 2.9. The approximating set $A_{(2)}$ for the cross, with initial set as above.

2.3.4.2 A decoration

The IFS now consists of the affine mappings f_j $(1 \le j \le 4)$ given by

$$L_1 = \begin{pmatrix} \frac{1}{3} & 0 \\ 0 & \frac{1}{3} \end{pmatrix}, \qquad b_1 = \begin{pmatrix} \frac{1}{3} \\ 0 \end{pmatrix},$$

$$L_2 = \begin{pmatrix} 0 & -\frac{1}{3} \\ \frac{1}{3} & 0 \end{pmatrix}, \qquad b_2 = \begin{pmatrix} \frac{1}{3} \\ 0 \end{pmatrix},$$

$$L_3 = \begin{pmatrix} 0 & \frac{1}{3} \\ -\frac{1}{3} & 0 \end{pmatrix}, \qquad b_3 = \begin{pmatrix} \frac{2}{3} \\ \frac{1}{3} \end{pmatrix},$$

$$L_4 = \begin{pmatrix} \frac{1}{3} & 0 \\ 0 & \frac{2}{3} \end{pmatrix}, \qquad b_4 = \begin{pmatrix} \frac{1}{3} \\ \frac{1}{3} \end{pmatrix}.$$

Using the half of the unit square below the diagonal as initial set $A_{(0)}$ one sees (Figure 2.10) that f_1, f_2 and f_3 are similarities with similarity factor $\frac{1}{3}$, but f_2 turns the image a right angle counterclockwise while f_3 turns it a right angle clockwise. The mapping f_4 is not any more a similarity since it squeezes everything in direction of the x-axis by a factor $\frac{1}{3}$, but in direction of the y-axis by a factor of $\frac{2}{3}$. Although the open set condition is satisfied, Theorem 1.3.8 is no longer applicable. It is still possible to compute the box-counting dimension of the resulting fractal (Figure 2.11). A closer inspection, using as $A_{(0)}$ the boundary of the unit square (Figure 2.12), reveals that the set $A_{(k)}$ is minimally covered by 5^k lattice squares of side length 3^{-k}. By Theorem 1.2.2 we get again $\dim_B(A) = \frac{\log 5}{\log 3} \approx 1.465$.

Figure 2.10. The generating set $A_{(1)}$ for the decoration, consisting of four images of the initial set $A_{(0)}$ which consists of the half of the unit square below the diagonal off the origin.

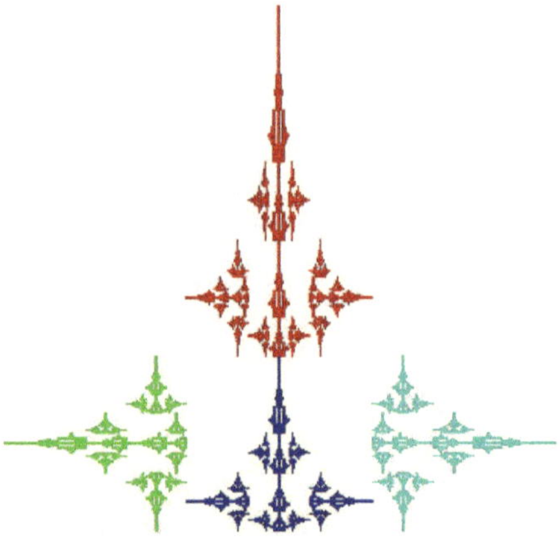

Figure 2.11. The approximating set $A_{(7)}$ for the decoration, with initial set the boundary of the unit square.

Figure 2.12. The approximating set $A_{(3)}$ for the decoration, with initial set $A_{(0)}$ the boundary of the unit square.

2.3.4.3 An antenna

The IFS consists of the affine mappings f_j $(1 \le j \le 4)$ given by

$$L_1 = \begin{pmatrix} \frac{1}{3} & 0 \\ 0 & \frac{1}{3} \end{pmatrix}, \qquad b_1 = \begin{pmatrix} \frac{1}{3} \\ 0 \end{pmatrix},$$

$$L_2 = \begin{pmatrix} 0 & \frac{1}{3} \\ -\frac{1}{3} & 0 \end{pmatrix}, \qquad b_2 = \begin{pmatrix} \frac{1}{3} \\ \frac{1}{3} \end{pmatrix},$$

$$L_3 = \begin{pmatrix} 0 & -\frac{1}{3} \\ \frac{1}{3} & 0 \end{pmatrix}, \qquad b_3 = \begin{pmatrix} \frac{2}{3} \\ \frac{1}{3} \end{pmatrix},$$

$$L_4 = \begin{pmatrix} \frac{1}{3} & 0 \\ 0 & \frac{2}{3} \end{pmatrix}, \qquad b_4 = \begin{pmatrix} \frac{1}{3} \\ \frac{1}{3} \end{pmatrix}.$$

There is too much overlap of the images of the unit square as initial set (Figure 2.13) to hope for an open set condition, and f_4 is certainly not a similarity. In order not to get involved in too much complication it will suffice to notice that the approximating set $A_{(k)}$ may be covered by less than 5^k lattice squares of side length 3^{-k} (Figure 2.14a). The dimension of the resulting fractal will therefore not exceed $\frac{\log 5}{\log 3} \approx 1.465$. Starting from the boundary $A_{(0)}$ of the unit square the set $A_{(1)}$ looks simple, but no longer the set $A_{(7)}$ approximating the fractal A which resembles an intricate TV-antenna (Figure 2.14b).

The data for the next fractals to be presented here have been taken from or have been at least inspired by examples in [Barnsley, 1988] and [Peitgen, Jürgens, Saupe, 1992].

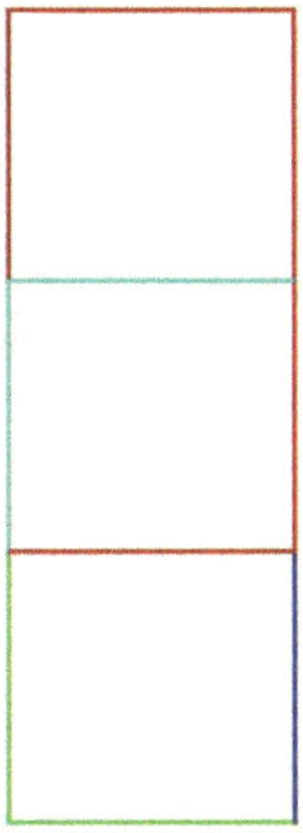

Figure 2.13. The approximating set $A_{(1)}$ for the antenna, consisting of four overlapping images of the initial set $A_{(0)}$ which consists of the boundary of the unit square.

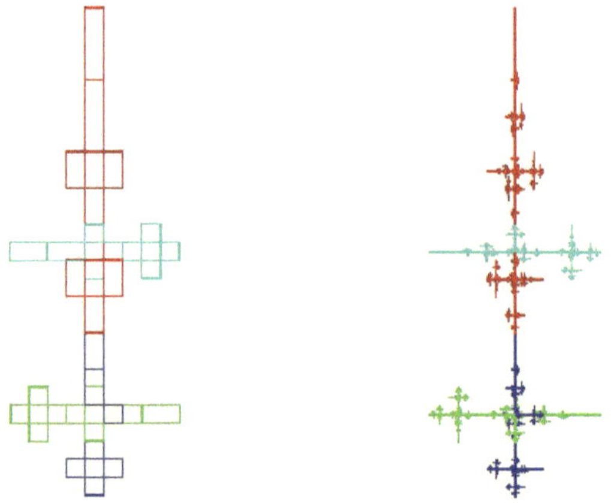

(a) (b)

Figure 2.14. The approximating sets $A_{(3)}$ (Figure 2.14a) and $A_{(7)}$ (b) for the antenna, with initial set as above.

2.3.4.4 A flower garden

Consider the three affine mappings defined as follows (Figure 2.15, cf. Figure 5.12 in [Peitgen, Jürgens, Saupe, 1992]):

$$L_1 = \begin{pmatrix} \frac{1}{3} & 0 \\ 0 & \frac{1}{3} \end{pmatrix}, \qquad b_1 = \begin{pmatrix} \frac{1}{3} \\ \frac{2}{3} \end{pmatrix},$$

$$L_2 = \begin{pmatrix} 0 & -\frac{1}{3} \\ 1 & 0 \end{pmatrix}, \qquad b_2 = \begin{pmatrix} \frac{1}{3} \\ 0 \end{pmatrix},$$

$$L_3 = \begin{pmatrix} 0 & \frac{1}{3} \\ -1 & 0 \end{pmatrix}, \qquad b_3 = \begin{pmatrix} \frac{2}{3} \\ 1 \end{pmatrix}.$$

At a first glance there seems to be some confusion: the mappings f_2 and f_3 are no contractions. Still, looking at the set $A_{(2)}$ (obtained just as the set $A_{(1)}$ from the boundary of the unit square as initial set) everything has indeed been contracted (Figure 2.16). Why? The reason becomes clearer if as initial set $A_{(0)}$ we take a triangle below the diagonal of the unit square (Figure 2.17): the mapping f_2 not only squeezes everything to $\frac{1}{3}$ its breadth but does so only after having turned everything about a right angle counterclockwise; similarly the mapping f_3 squeezes everything to $\frac{1}{3}$ its breadth after having it turned about a right angle clockwise. This causes the square of the collage $F = \bigcup_{j=1}^{3} f_j$ to be a contraction and its fixed point A to exist as an attractor (Figure 2.18).

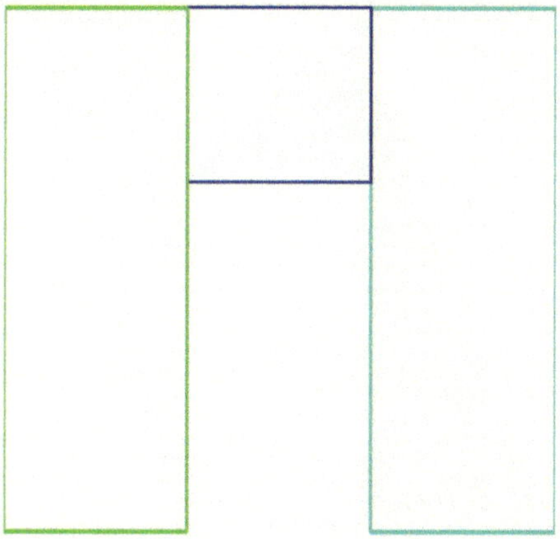

Figure 2.15. The generating set $A_{(1)}$ for the flower garden, starting with the boundary of the unit square as initial set $A_{(0)}$.

Figure 2.16. The approximating set $A_{(2)}$, again starting with the boundary of the unit square as initial set $A_{(0)}$.

Figure 2.17. The generating set $A_{(1)}$ for the flower garden, starting with half the unit square below the diagonal as initial set $A_{(0)}$.

Figure 2.18. The approximating set $A_{(11)}$ for the flower garden, starting with either of the mentioned initial sets.

If the mappings f_2 and f_3 would have been replaced by the mappings f_2' and f_3' given e.g. by

$$L_2' = \begin{pmatrix} 1 & 0 \\ 0 & \frac{1}{3} \end{pmatrix}, \qquad b_2 = \begin{pmatrix} 0 \\ 0 \end{pmatrix},$$

$$L_3' = \begin{pmatrix} 1 & 0 \\ 0 & \frac{1}{3} \end{pmatrix}, \qquad b_3 = \begin{pmatrix} \frac{2}{3} \\ 0 \end{pmatrix},$$

then the set $A'_{(1)}$ (starting with the boundary of the unit square as $A'_{(0)}$) would again have looked as in Figure 2.15 and the sequence $\{F^{(k)}(A'_{(0)})\}_{k=1}^{\infty}$ would have converged in the HAUSDORFF metric to a limiting set A', but the limiting set would depend on the initial set: taking e.g. $A_{(0)}$ in the unit square disjoint from its lower half would produce a limiting set A' again disjoint from this lower half.

Again a closer inspection of the sets $A_{(k)}$ reveals that they are minimally covered by 7^k lattice squares of side length 3^{-k}. The dimension of the fractal A (enforcing some fantasy one might think of it as the ground map of some flower garden) by Theorem 1.2.2 is $\dim_B = \frac{\log 7}{\log 3} \approx 1.771$.

2.3.4.5 A pentagon snowflake

In order to define a parameter family of IFS let us start out with a disc $A_{(0)}$ with center $m = (\frac{1}{2}, \frac{1}{2})$ and radius $\frac{1}{2}$, inscribed into the unit square, and five smaller discs $A_{1,j}$ $(1 \le j \le 5)$ with centers m_j and radius r, tangentially touching their neighbours and the boundary of $A_{(0)}$ (Figure 2.19). For $d = |m - m_j|$ and $\alpha = \frac{\pi}{5} \sim 36°$ the equations

$$r + d = \frac{1}{2}, \qquad \frac{r}{d} = \sin\alpha$$

give us

$$\begin{aligned} r &= \frac{\sin\alpha}{2(1+\sin\alpha)} \approx 0.1851, \\ d &= \frac{1}{2(1+\sin\alpha)} \approx 0.3149, \\ m_j &= \begin{pmatrix} \frac{1}{2} - d\cdot\sin 2(j+1)\alpha \\ \frac{1}{2} + d\cdot\cos 2(j+1)\alpha \end{pmatrix} \qquad (1 \le j \le 5). \end{aligned}$$

Using the parameter $t \in [1, \frac{d}{r}]$ we define f_j to be the similarity with similarity factor $ts = 2tr \approx 0.3702\,t$ mapping $A_{(0)}$ onto the disc with center m_j and radius tr, given as in (2.4) by

$$L_j = t\cdot\begin{pmatrix} s & 0 \\ 0 & s \end{pmatrix} \qquad (1 \le j \le 5),$$

$$b_j = \begin{pmatrix} \frac{1}{2} - d\cdot\sin 2(j+1)\alpha - \frac{ts}{2} \\ \frac{1}{2} + d\cdot\cos 2(j+1)\alpha - \frac{ts}{2} \end{pmatrix} \qquad (1 \le j \le 5).$$

Figure 2.19. The generating set $(A_1)_{(1)}$ for the pentagon snowflake, starting with a disc of radius 0.5, inscribed in the unit square as initial set $A_{(0)}$.

For $t = 1$ the open set condition is satisfied and the dimension of the resulting fractal A (Figure 2.20) by Theorem 1.3.8 is $\dim_H(A) = \frac{\log 5}{-\log s} \approx 1.62$.

As the variable t increases, the similarity factor for the mappings f_j increases to ts and the radius of the five discs constituting $A_{(1)}$ increases to $\frac{ts}{2}$, but their centers m_j remain fixed. The attractor A_t changes correspondingly in a continuous way as already stated in Theorem 2.2.12. The open set condition is no longer satisfied from the beginning, but as long as the five sets $f_j(A_t)$ lie apart from each other we can still apply our heuristic reasoning of Section 1.1.3 to obtain $\dim_S(A_t) = \frac{\log 5}{-\log t - \log s}$. In fact, for $t \approx 1.0319$ the fractal becomes a connected set and one obtains $ts \approx 0.382$, $\dim_S(A_t) \approx 1.672$ (Figure 2.21, cf. Figure 5.15 in [Peitgen, Jürgens, Saupe, 1992]).

Figure 2.20. The approximating set $(A_1)_{(9)}$ for the pentagon snowflake with similarity factor $s \approx 0.37019$, $\dim_H(A_1) \approx 1.62$.

Figure 2.21. The approximating set $(A_t)_{(9)}$ for pentagon snowflake, obtained with similarity factor $ts \approx 0.382$, $\dim_S(A_t) \approx 1.672$.

2.3.4.6 Evolution of a triplicate continent

Let us modify the setup of Section 2.3.4.5 by replacing the five smaller discs by only three discs A_j $(1 \le j \le 3)$ with centers m_j and radius r, again tangentially touching their neighbours and the boundary of $A_{(0)}$ (Figure 2.22). For $d = |m - m_j|$ and $\alpha = \frac{\pi}{3} \sim 60°$, the same equations as in Section 2.3.4.5 give us

$$r = \frac{\sin\alpha}{2(1+\sin\alpha)} = \frac{\sqrt{3}}{2(2+\sqrt{3})} \approx 0.2321,$$

$$d = \frac{1}{2(1+\sin\alpha)} \approx 0.2679,$$

$$m_j = \begin{pmatrix} \frac{1}{2} - d\cdot\sin 2j\alpha \\ \frac{1}{2} + d\cdot\cos 2j\alpha \end{pmatrix} \qquad (1 \le j \le 3).$$

Using the parameter $t \in [1, \frac{d}{r}]$ we now define $f_{t,j}$ to be the similarity with similarity factor $ts = 2tr \approx 0.4642\,t$ mapping $A_{(0)}$ onto the disc with center m_j and radius tr, given as in (2.4) by

$$L_j = t\cdot\begin{pmatrix} 0 & s \\ -s & 0 \end{pmatrix},$$

$$b_j = \begin{pmatrix} \frac{1}{2} - d\cdot\sin 2j\alpha - \frac{ts}{2} \\ \frac{1}{2} + d\cdot\cos 2j\alpha + \frac{ts}{2} \end{pmatrix} \qquad (1 \le j \le 3).$$

Again, for $t = 1$, the open set condition is satisfied and the dimension of the resulting fractal A_1 by Theorem 1.3.8 is $\dim_H(A_1) = \frac{\log 3}{-\log s} \approx 1.4311$. Still, every mapping $f_{t,j}$ also turns the image about a right angle clockwise. As a result the fractal A_1 looks like

Figure 2.22. The set $(A_1)_{(1)}$ for the island archipelago if the initial set $A_{(0)}$ is a disc of radius 0.5 inscribed in the unit square.

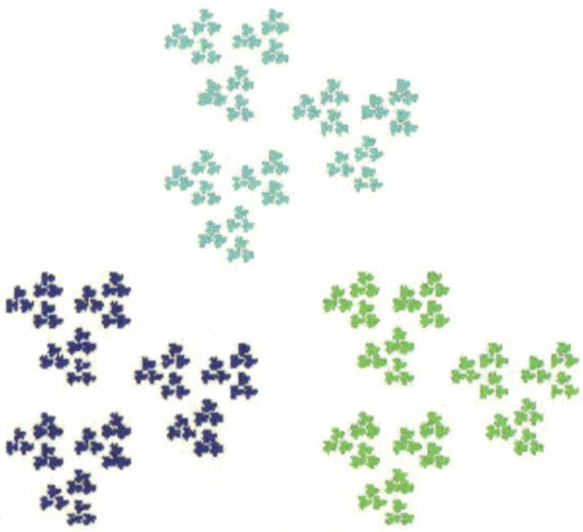

Figure 2.23. The approximating set $(A_1)_{(9)}$ for the island archipelago A_1 corresponding to the similarity factor $s \approx 0.4642$, with $\dim_H(A_1) \approx 1.4311$.

a triplicate spiraling archipelago of tiny islands (Figure 2.23) which, in fact, are tiny archipelagoes in their own right.

What happens if we let the parameter t increase? The similarity factor increases (Figure 2.24), and with it the size of the archipelagoes, but again, by the construction of the mappings, the images $f_{t,j}(\frac{1}{2}, \frac{1}{2})$ of the center of $A_{(0)}$ remain constant. The open set condition evaporates in the mist, but since the archipelagoes at the beginning are lying apart one might feel encouraged to stick to formula (1.7) providing $\dim_S(A_t) = \frac{\log 3}{-\log t - \log s}$.

Figure 2.24. The generating set $(A_t)_{(1)}$ for the island archipelago if the initial set $(A_t)_{(0)} = A_{(0)}$ is a disc of radius 0.5 inscribed in the unit square, and $t = 1.1$.

As t increases so does $\dim_S(A_t)$, and whatever mass might be concentrated in the islands seems to drift together (Figures 2.25, 2.26), until for $t = \frac{2+\sqrt{3}}{3} \approx 1.244$ one gets $\frac{\log 3}{-\log t - \log s} = 2$ and all archipelagoes are joined in a new continent consisting of three countries similar in size to the whole continent (Figure 2.27, cf. Figure 5.11 in [Peitgen, Jürgens, Saupe, 1992]).

Figure 2.25. The approximating set $(A_t)_{(9)}$ for the island archipelago A_t corresponding to $t = 1.1$ and similarity factor $ts \approx 0.51051$, with $\dim_S(A_r) \approx 1.6340$.

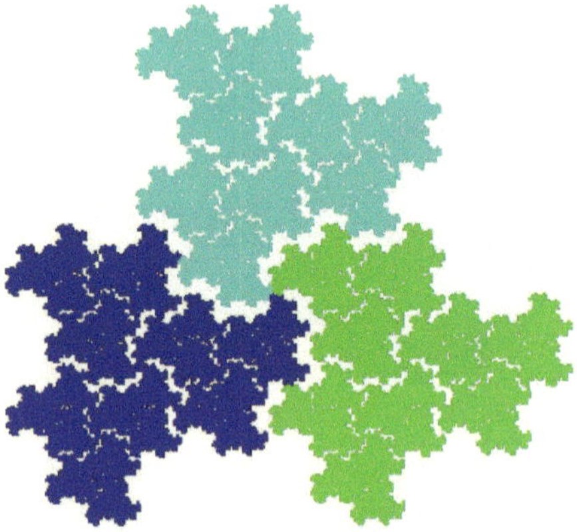

Figure 2.26. The approximating set $(A_t)_{(9)}$ for the island archipelago A_t corresponding to $t = 1.2$ and similarity factor $ts \approx 0.55692$, with $\dim_S(A_t) \approx 1.8769$.

Figure 2.27. The approximating set $(A_t)_{(9)}$ for the triplicate continent A_t corresponding to $t = \frac{2+\sqrt{3}}{3} \approx 1.2440$, with $\dim_S(A_t) = 2$.

2.3.4.7 A twig

While the fractals presented so far originate from geometric objects, it is also possible to imitate nature. Or is nature governed by hidden geometric laws?

One map (f_3) of the next IFS is almost degenerate and helps to draw some sticks (Figures 2.28 and 2.29, cf. Figure 5.13 in [Peitgen, Jürgens, Saupe, 1992]):

$$L_1 = \begin{pmatrix} 0.387 & 0.430 \\ 0.430 & -0.387 \end{pmatrix}, \qquad b_1 = \begin{pmatrix} 0.176 \\ 0.522 \end{pmatrix},$$

$$L_2 = \begin{pmatrix} 0.441 & -0.091 \\ -0.009 & -0.322 \end{pmatrix}, \qquad b_2 = \begin{pmatrix} 0.342 \\ 0.506 \end{pmatrix},$$

$$L_3 = \begin{pmatrix} -0.468 & 0.020 \\ -0.113 & 0.015 \end{pmatrix}, \qquad b_3 = \begin{pmatrix} 0.320 \\ 0.400 \end{pmatrix}.$$

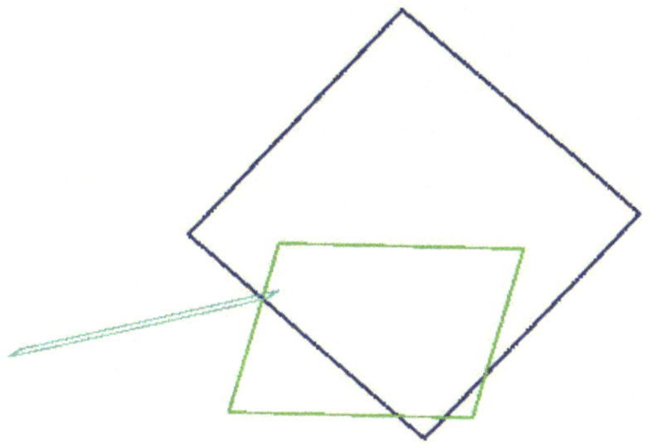

Figure 2.28. The generating set $A_{(1)}$ for the twig, if $A_{(0)}$ is the boundary of the unit square.

Figure 2.29. The approximating set $A_{(10)}$ for the twig.

2.3.4.8 A tree (cf. Figure 5.16 in [Peitgen, Jürgens, Saupe, 1992])

$$L_1 = \begin{pmatrix} 0.195 & -0.488 \\ 0.344 & 0.443 \end{pmatrix}, \qquad b_1 = \begin{pmatrix} 0.443 \\ 0.245 \end{pmatrix},$$

$$L_2 = \begin{pmatrix} 0.462 & 0.414 \\ -0.252 & 0.361 \end{pmatrix}, \qquad b_2 = \begin{pmatrix} 0.251 \\ 0.569 \end{pmatrix},$$

$$L_3 = \begin{pmatrix} -0.058 & -0.070 \\ 0.453 & -0.111 \end{pmatrix}, \qquad b_3 = \begin{pmatrix} 0.598 \\ 0.097 \end{pmatrix},$$

$$L_4 = \begin{pmatrix} -0.035 & 0.070 \\ -0.469 & -0.022 \end{pmatrix}, \qquad b_4 = \begin{pmatrix} 0.488 \\ 0.507 \end{pmatrix},$$

$$L_5 = \begin{pmatrix} -0.637 & 0 \\ 0 & 0.501 \end{pmatrix}, \qquad b_5 = \begin{pmatrix} 0.856 \\ 0.251 \end{pmatrix}.$$

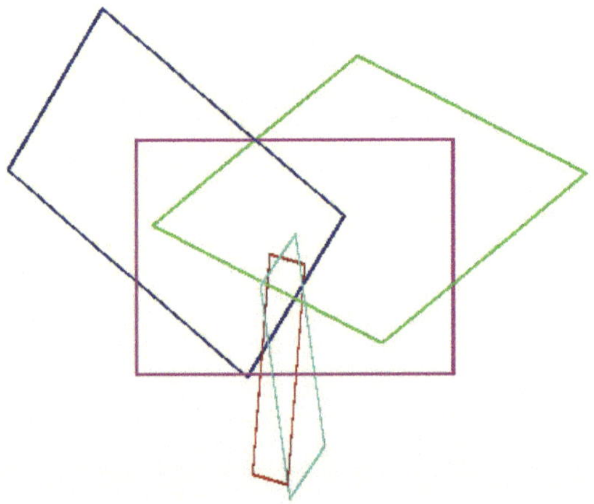

Figure 2.30. The generating set $A_{(1)}$ for the tree, if $A_{(0)}$ is the boundary of the unit square.

Figure 2.31. The approximating set $A_{(7)}$ for the tree.

2.3.4.9 A leaf

The IFS here consists of four similarities (cf. Figure 3.10.10 in [Barnsley, 1988]):

$$L_1 = L_2 = \begin{pmatrix} 0.6 & -0 \\ 0 & 0.6 \end{pmatrix}, \qquad b_1 = \begin{pmatrix} 0.18 \\ 0.36 \end{pmatrix}, \quad b_2 = \begin{pmatrix} 0.18 \\ 0.12 \end{pmatrix},$$

$$L_3 = \begin{pmatrix} 0.4 & 0.3 \\ -0.3 & 0.4 \end{pmatrix}, \qquad b_3 = \begin{pmatrix} 0.27 \\ 0.36 \end{pmatrix},$$

$$L_4 = \begin{pmatrix} 0.4 & -0.3 \\ 0.3 & 0.4 \end{pmatrix}, \qquad b_4 = \begin{pmatrix} 0.27 \\ 0.09 \end{pmatrix}.$$

It is instructive to study in more detail a spectacular fractal.

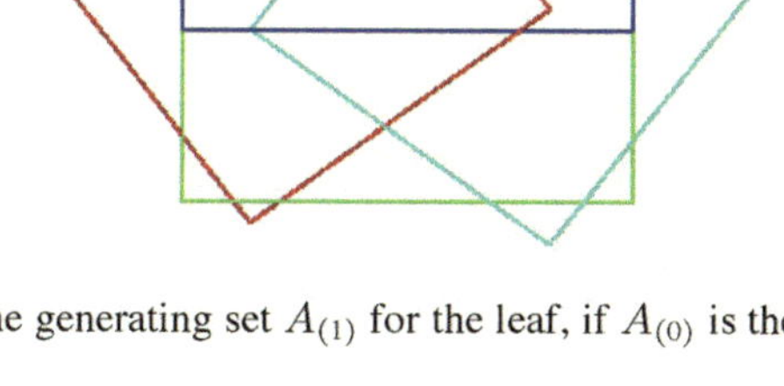

Figure 2.32. The generating set $A_{(1)}$ for the leaf, if $A_{(0)}$ is the boundary of the unit square.

Figure 2.33. The approximating set $A_{(9)}$ for the leaf.

2.3.4.10 The BARNSLEY fern

The IFS consists of four affine contractions f_j $(1 \leq j \leq 4)$, f_2 and f_3 being similarities (Figure 2.34, cf. Figure 5.25 in [Peitgen, Jürgens, Saupe, 1992] and Table 3.8.3 in [Barnsley, 1988]):

$$L_1 = \begin{pmatrix} 0 & 0 \\ 0 & 0.160 \end{pmatrix}, \qquad b_1 = \begin{pmatrix} 0.500 \\ 0 \end{pmatrix},$$

$$L_2 = \begin{pmatrix} 0.849 & 0.037 \\ -0.037 & 0.849 \end{pmatrix}, \qquad b_2 = \begin{pmatrix} 0.075 \\ 0.183 \end{pmatrix},$$

$$L_3 = \begin{pmatrix} 0.197 & -0.226 \\ 0.226 & 0.197 \end{pmatrix}, \qquad b_3 = \begin{pmatrix} 0.400 \\ 0.049 \end{pmatrix},$$

$$L_4 = \begin{pmatrix} -0.15 & 0.283 \\ 0.260 & 0.237 \end{pmatrix}, \qquad b_4 = \begin{pmatrix} 0.575 \\ -0.084 \end{pmatrix}.$$

Looking at the resulting fractal A (Figure 2.35) one wonders how this comes about. It may help to study the individual mappings f_j more in detail (we round numerical results to two decimals). f_1 maps the whole unit square Q with contraction constant $c_1 = 0.16$ and fixed point $(0.5, 0)$ onto the interval $I_u = [0, 0.16]$ on the y-parallel symmetry-axis u of Q. Every time f_1 is applied to a subset of $\mathbb{R}^2$ this set is projected orthogonally upon u and reduced to 0.16 of its original height; in particular, a set "reaching from height 0 to height 1" has as its image the interval I_u on u. The mapping f_1 thus provides for the bottom stem of the fern. The map f_2 has determinant (=area reducing factor) 0.72, contraction constant $c_2 = 0.85$ and fixed point $z_2 = (0.75, 1.03)$ (note that the matrices L_2 and L_3 are similarity matrices i.e. constant multiples of orthogonal matrices). It takes the unit square, reduces it to 0.85 of its size (and therefore to $0.72 \approx 0.85^2$ of its area), turns it a little bit down to the right and moves it with the midpoint of its lower side up into the upper endpoint of the interval I_u. Every time it is applied to something contained in Q it takes this something along on its journey towards the fixed point near the upper side of Q, reducing and turning and moving it, and providing so for the typical shrinking of the leaves close to the top of the fern in the fixed point of f_2. The map f_3 has determinant 0.09, contraction constant $c_3 = 0.30$ and fixed point $(0.45, 0.19)$. It maps Q into a square $f_3(Q)$ with vertices $A_3 = (0.40, 0.05)$,

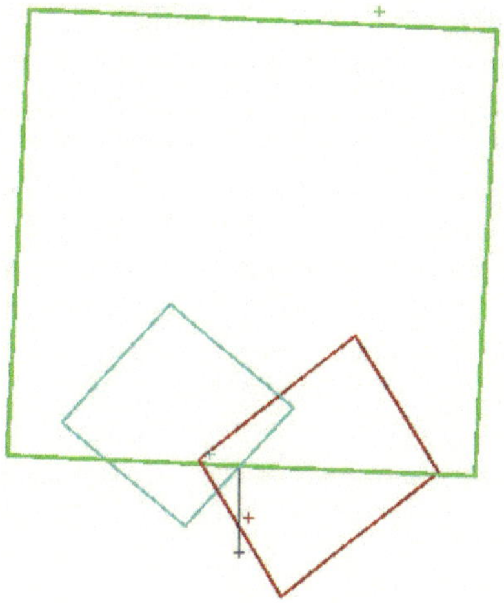

Figure 2.34. The generating set $A_{(1)}$ for the BARNSLEY fern, if $A_{(0)}$ is the boundary of the unit square.

Figure 2.35. The approximating set $A_{(29)}$ for the BARNSLEY fern.

$B_3 = (0.60, 0.28)$, $C_3 = (0.37, 0.58)$, $D_3 = (0.17, 0.38)$. So whatever happens in Q is going to be similarly happening in $f_3(Q)$, reduced in size by the factor 0.30, and turned and moved along from Q to $f_3(Q)$, and as "time" k passes by, being contracted into z_3. This provides for the leaves to be similar images of the whole fern and hanging on to the left of it. Finally, f_4 has determinant -0.11 (the minus due to change in orientation), contraction constant $c_4 = 0.38$ and fixed point $z_4 = (0.52, 0.07)$. In a similar way as f_3 it provides for the leaves to the right of the fern, which are, however, not similar but only affine images of the whole fern, and while the left leaves follow the turning of the whole fern to the right, the right leaves now turn to the left.

It is also instructive to observe the action of the IFS upon a polygon ∂E (the symbol ∂ denoting the boundary of the set E) roughly shaped as the contour of the limiting fractal A, such as e.g. the quadrangle with vertices $a = (0.50, 0)$, $b = (0.75, 0.30)$, $c = (0.70, 1.05)$, $d = (0.30, 0.40)$ (Figure 2.36). What happens is that (if E is the interior of the quadrangle $abcd$) the HAUSDORFF distance $h(E, F(E))$ becomes small (Figure 2.37). This is not surprising, since if E is close to A then so is $F(E)$.

Figure 2.36. A quadrangle E approximating the BARNSLEY fern.

Figures 2.38, 2.39 and 2.40 show how the structure of $F^{(k)}(E)$ develops as k increases: the region towards the top inherits more and more of the lower part, while the lower branches profit more and more from the entire structure.

The relation between the sets E and $F(E)$ just mentioned above works the other way too: if $h(E, F(E))$ is small, then by the Collage Theorem 2.2.11 the set E must be close to the limiting fractal A. This opens the possibility to construct a fractal "looking approximately like a given set A". The recipe is as follows: construct (a) a polygon E approximating as well as desired – in the HAUSDORFF metric – the prospective fractal set A; construct (b) affine contractions f_j $(1 \leq j \leq J)$ with (c) contraction constants as small as desired (all of them less than $c < 1$, say) such that (d) the set E is approximated as well as desired by $F(E) := \bigcup_{j=1}^{J} f_j(E)$. Then by the Collage Theorem 2.2.11 for the fractal A_F, the fixed point of F, we get $h(A, A_F) \leq h(A, E) + h(E, A_F) \leq$

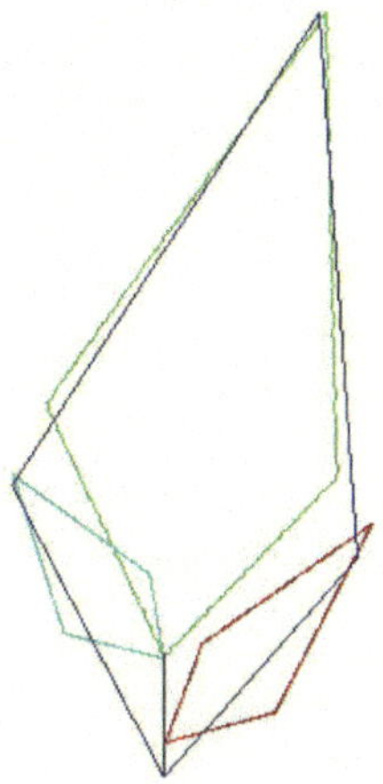

Figure 2.37. The generating set $E_{(1)}$ for the BARNSLEY fern, and the quadrangle E used as initial set.

Figure 2.38. The approximating set $E_{(2)}$ for the BARNSLEY fern, if $E_{(0)}$ is the quadrangle E.

Figure 2.39. The approximating set $E_{(4)}$ for the BARNSLEY fern, if $E_{(0)}$ is the quadrangle E.

Figure 2.40. The approximating set $E_{(8)}$ for the BARNSLEY fern, if $E_{(0)}$ is the quadrangle E.

$h(A, E) + \frac{1}{1-c} h(E, F(E))$, which then becomes as small as desired.

Of course we also need a recipe for how to satisfy all these desires. Problem (a) does not offer any special difficulties. The problems (b), (c) and (d) may be tackled with the help of the following theorem.

2.3.5 Theorem. *Let $p_i = (p_{i,1}, p_{i,2})$ $(1 \le i \le 3)$ be the vertices of a non-degenerated triangle. Given any three points $q_i = (q_{i,1}, q_{i,2})$ there is an affine map f satisfying $f(p_i) = q_i$ $(1 \le i \le 3)$.*

Proof. We look for a 2×2-matrix L and a vector $b \in \mathbb{R}^2$,

$$L = \begin{pmatrix} l_{11} & l_{12} \\ l_{21} & l_{22} \end{pmatrix}, \qquad b = \begin{pmatrix} b_1 \\ b_2 \end{pmatrix},$$

satisfying

$$q_{i,j} = l_{j,1} p_{i,1} + l_{j,2} p_{i,2} + b_j \qquad (1 \le i \le 3,\ j = 1, 2).$$

This provides for two systems of three inhomogeneous linear equations each for the unknown variables $(l_{j,1}, l_{j,2}, b_j)$. It is uniquely solvable if (bars $|\cdot|$ and "det" denoting, as usual, determinants)

$$\begin{aligned} 0 \ne \begin{vmatrix} p_{1,1} & p_{1,2} & 1 \\ p_{2,1} & p_{2,2} & 1 \\ p_{3,1} & p_{3,2} & 1 \end{vmatrix} &= \begin{vmatrix} p_{1,1} & p_{1,2} & 1 \\ p_{2,1} - p_{1,1} & p_{2,2} - p_{1,2} & 0 \\ p_{3,1} - p_{1,1} & p_{3,2} - p_{1,2} & 0 \end{vmatrix} \\ &= \begin{vmatrix} p_{2,1} - p_{1,1} & p_{2,2} - p_{1,2} \\ p_{3,1} - p_{1,1} & p_{3,2} - p_{1,2} \end{vmatrix} = \det(p_2 - p_1, p_3 - p_1). \end{aligned}$$

This condition is satisfied since the triangle p_1, p_2, p_3 is supposed to be non-degenerate. □

2.3.6 Construction of a blossom fractal

Let us try an example. Suppose we want to picture by means of a fractal a blossom with five leaflets fitting roughly into the polygon with vertices $a = (0.50, 0.30)$, $b = (0.70, 0.40)$, $c = (0.65, 0.60)$, $d = (0.45, 0.55)$, $e = (0.35, 0.40)$ (Figure 2.41a). To this end we cover the region in the interior P of the polygon, which is supposed to display eventually the desired fractal, by five affine images of P (Figure 2.42a).

How have we arrived at these images? We have chosen the images of the triangle ace in a suitable way (Figure 2.42b); this choice requires a little bit of geometric intuition, but everything not to our taste may simply be discarded and replaced by another triangle $a_j c_j e_j$ until we are satisfied the result complies with requirement (d). These

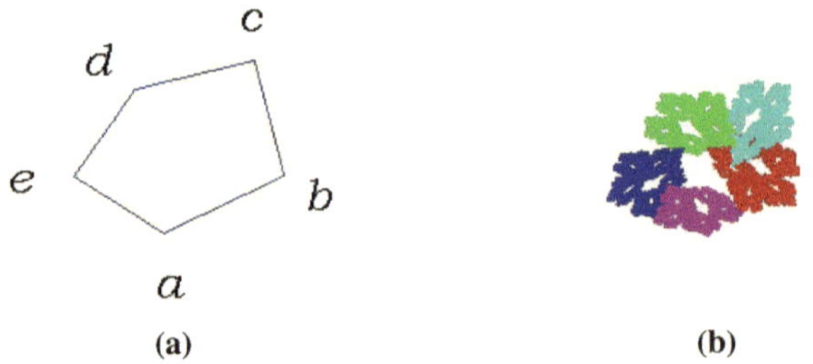

Figure 2.41. A pentagon *abcde* (2.41a) as initial set for a prospective blossom (2.41b).

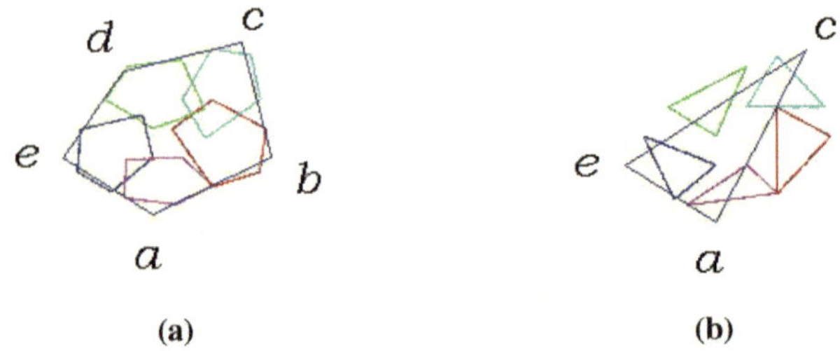

Figure 2.42. The pentagon *abcde* (2.42a) and the triangle *ace* (2.42b), each with its five images under the affine mappings f_j $(1 \le j \le 5)$.

choices require the affine mappings f_j $(1 \le j \le 5)$ to furnish the following images:

$$
\begin{array}{llll}
f_1: & a_1 = (0.50, 0.40), & c_1 = (0.38, 0.45), & e_1 = (0.43, 0.34), \\
f_2: & a_2 = (0.50, 0.45), & c_2 = (0.55, 0.57), & e_2 = (0.42, 0.50), \\
f_3: & a_3 = (0.55, 0.50), & c_3 = (0.68, 0.50), & e_3 = (0.60, 0.59), \\
f_4: & a_4 = (0.60, 0.50), & c_4 = (0.60, 0.35), & e_4 = (0.69, 0.45), \\
f_5: & a_5 = (0.55, 0.40), & c_5 = (0.45, 0.33), & e_5 = (0.60, 0.35).
\end{array}
$$

This produces mappings f_j $(1 \le j \le 5)$ given by the following matrices L_j and translation vectors b_j, with fixed points zf_j, determinants df_j and contraction constants

cf_j (all entries rounded to the given decimals):

j	L_j	b_j	zf_j	df_j	cf_j
1	$\begin{pmatrix} 0.15 & -0.48 \\ 0.38 & -0.02 \end{pmatrix}$	$\begin{pmatrix} 0.57 \\ 0.22 \end{pmatrix}$	$\begin{pmatrix} 0.455 \\ 0.381 \end{pmatrix}$	0.18	0.53
2	$\begin{pmatrix} 0.48 & -0.08 \\ 0.05 & 0.43 \end{pmatrix}$	$\begin{pmatrix} 0.28 \\ 0.35 \end{pmatrix}$	$\begin{pmatrix} 0.462 \\ 0.564 \end{pmatrix}$	0.20	0.52
3	$\begin{pmatrix} 0.03 & 0.45 \\ 0.45 & 0.22 \end{pmatrix}$	$\begin{pmatrix} 0.43 \\ 0.66 \end{pmatrix}$	$\begin{pmatrix} 0.628 \\ 0.484 \end{pmatrix}$	0.19	0.59
4	$\begin{pmatrix} 0.45 & 0.22 \\ 0.00 & -0.50 \end{pmatrix}$	$\begin{pmatrix} 0.76 \\ 0.65 \end{pmatrix}$	$\begin{pmatrix} 0.590 \\ 0.433 \end{pmatrix}$	0.23	0.60
5	$\begin{pmatrix} 0.42 & -0.13 \\ 0.13 & -0.30 \end{pmatrix}$	$\begin{pmatrix} 0.80 \\ 0.42 \end{pmatrix}$	$\begin{pmatrix} 0.528 \\ 0.380 \end{pmatrix}$	0.14	0.44

The resulting fractal is illustrated in Figure 2.41b.

2.3.7 A moving grass (cf. Figure 3.11.3 in [Barnsley, 1988])

According to Theorem 2.2.12 a small change in the maps f_j of a given IFS will result in a small change of the attractor. In fact, it is possible to insert such a changing mechanism already in the definition of the IFS, as e.g. already done in the examples of Section 2.3.4.5 and Section 2.3.4.6. Let us conclude this section with the fractal image of a grass with a variable upper part.

$$L_1 = \begin{pmatrix} 0 & 0 \\ 0 & 0.3 \end{pmatrix}, \qquad b_1 = \begin{pmatrix} 0.5 \\ 0.11 \end{pmatrix},$$

$$L_2 = \begin{pmatrix} 0.60\cos\varphi - 0.032\sin\varphi & -0.698\sin\varphi \\ 0.032\cos\varphi + 0.60\sin\varphi & 0.698\cos\varphi \end{pmatrix},$$

$$b_2 = \begin{pmatrix} -0.3\cos\varphi + 0.1\sin\varphi + 0.5 \\ -0.04\cos\varphi - 0.24\sin\varphi + 0.3 \end{pmatrix},$$

$$L_3 = \begin{pmatrix} 0.322 & -0.414 \\ 0.278 & 0.414 \end{pmatrix}, \qquad b_3 = \begin{pmatrix} 0.404 \\ 0.008 \end{pmatrix},$$

$$L_4 = \begin{pmatrix} 0.246 & 0.437 \\ -0.223 & 0.349 \end{pmatrix}, \qquad b_4 = \begin{pmatrix} 0.312 \\ 0.368 \end{pmatrix}.$$

As φ increases from its initial value 0 the grass begins to bow in the wind until it finally breaks.

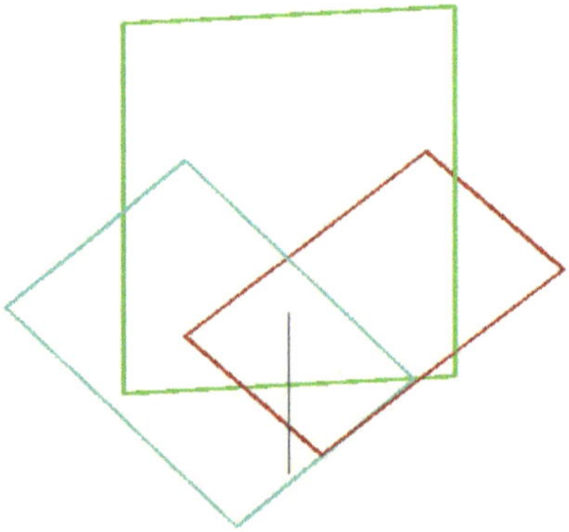

Figure 2.43. The generating set $A_{(1)}$ for the grass if the initial set is the boundary of the unit square.

Figure 2.44. The approximating set $A_{(12)}$ for the grass, with the boundary of the unit square as initial set.

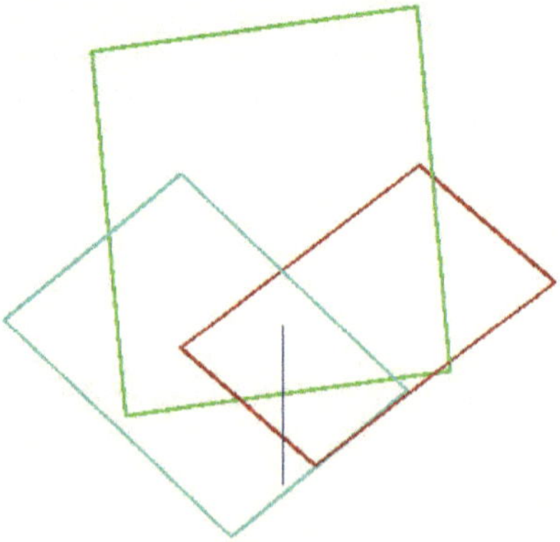

Figure 2.45. The generating set $A_{(1)}$ for the grass with an inclination angle $\varphi = 5°$ if the initial set is the boundary of the unit square.

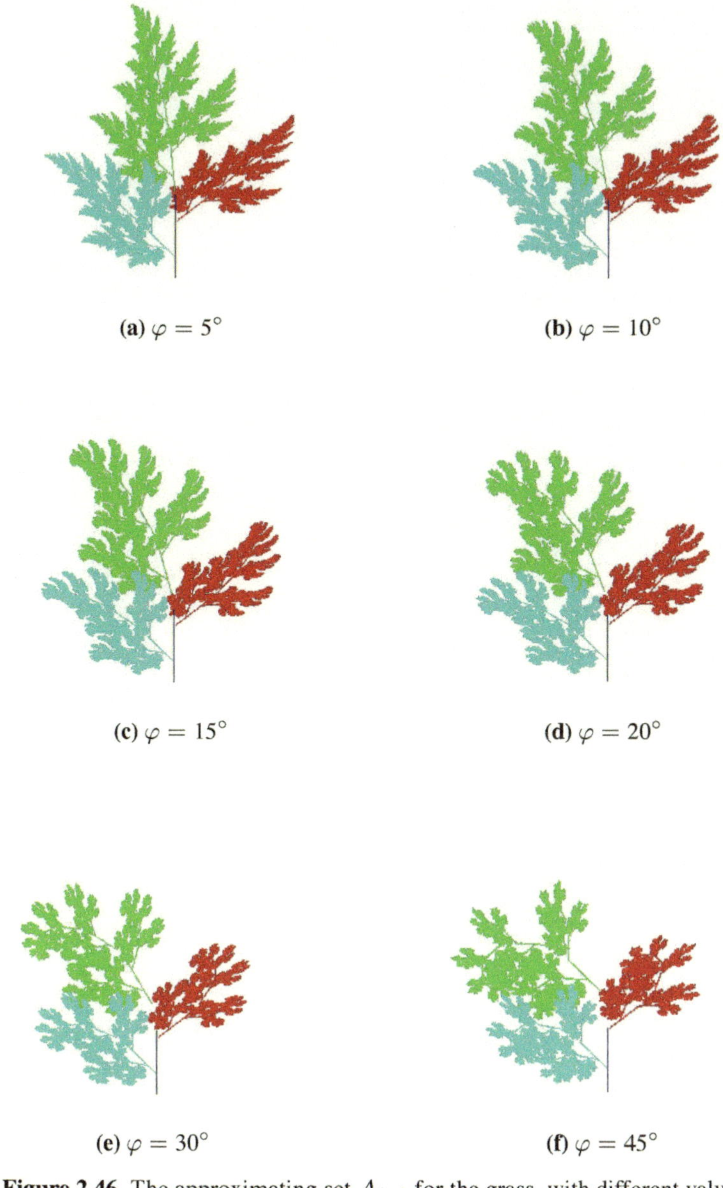

(a) $\varphi = 5^\circ$ **(b)** $\varphi = 10^\circ$

(c) $\varphi = 15^\circ$ **(d)** $\varphi = 20^\circ$

(e) $\varphi = 30^\circ$ **(f)** $\varphi = 45^\circ$

Figure 2.46. The approximating set $A_{(12)}$ for the grass, with different values of φ and the boundary of the unit square as initial set.

3 Iteration of complex polynomials

After having discussed the MENGER sponge we have already decided to stay in $\mathbb{R}^2$. We still have the possibility to identify $\mathbb{R}^2$ with the complex plane $\mathbb{C}$ and to make use of the algebraic operations defined for complex numbers $z = x + iy$. For instance we could use linear maps of the form $f(z) = cz + b$ ($c = c_1 + ic_2$, $b = b_1 + ib_2$, $c_j \in \mathbb{R}$, $b_j \in \mathbb{R}$). If we would limit ourselves to such maps we would lose a good part of our freedom to create fractals since all of these maps are similarities. If c is written as $c = |c|e^{i\gamma}$, then the map f turns everything in $\mathbb{C}$ about the angle γ (in radials) counterclockwise, multiplies the distances from the origin 0 by $|c|$ and finally moves the result a distance of $|b|$ in direction of the vector b. If $|c| < 1$, then f contracts with similarity factor $s = |c|$ and has the fixed point $z_f = \frac{b}{1-c}$. Every orbit $\{f^{(k)}(z)\}_{k=0}^{\infty}$ converges to z_f.

Something new happens if we decide to study orbits under more general maps than linear ones, and a natural next choice would be a polynomial of the form $f(z) = \sum_{j=0}^{n} c_j z^j$ ($c_j \in \mathbb{C}$, $0 \leq j \leq n$, $c_n \neq 0$). We have to think about only where to look for the equivalent of an attractor as in Chapter 2, which should be a set invariant under the action of f. Funnily enough it turns out that the interesting object connected with the iterations of a polynomial could be rather called a "repeller".

In order to get an impression of what will be waiting for us, let us consider a particularly simple example, namely the polynomial $f(z) = z^2$. This polynomial even has two fixed points: 0 and 1, the two solutions of the equation $f(z_f) = z_f$. But f behaves differently in the neighbourhood of 0 and 1. In general this behavior at a point z_0 is regulated by the value of the derivative of f, as becomes clear when a (small) complex increment h is added to z_0:

$$\begin{aligned} f(z_0 + h) &= f(z_0) + h \cdot f'(z_0) + o(h) \qquad \text{as } h \to 0, \\ |f(z_0 + h) - f(z_0)| &= |h| \cdot \left| f'(z_0) + \frac{o(h)}{h} \right| \begin{cases} < |h| & \text{if } |f'(z_0)| < 1, \\ > |h| & \text{if } |f'(z_0)| > 1. \end{cases} \end{aligned}$$

Taking $z_0 = z_f$ we get

$$|f(z_f + h) - z_f)| = |h| \cdot \left| f'(z_f) + \frac{o(h)}{h} \right| \begin{cases} < |h| & \text{if } |f'(z_f)| < 1, \\ > |h| & \text{if } |f'(z_f)| > 1. \end{cases}$$

Since $f'(z) = 2z$ we have $f'(0) = 0$ and $f'(1) = 2$. So $z_f = 0$ is an "attractive" fixed point, while $z_f = 1$ is a "repellent" one.

A generalization of a fixed point would be a point z_0 which the orbit $\{f^{(k)}(z_0)\}_{k=1}^{\infty}$ visits again: $f^{(p)}(z_0) = z_0$. Such a point is called *periodic* and the smallest positive integer p which does the trick is called the *period of* z_0. Looking for the periodic points of our map f we first note that $f^{(2)}(z) = z^4$, $f^{(3)}(z) = z^8$ and $f^{(k)}(z) = z^{2^k}$. So, apart from $z_0 = 0$, we have to look for points z_0 satisfying

$$z_0^{2^p} = z_0, \qquad z_0^{2^p - 1} = 1,$$

which implies $|z_0| = 1$. Writing $z_0 = e^{2\pi i\theta_0}$ we get

$$\left(e^{2\pi i\theta_0}\right)^{2^p-1} = e^{2\pi i\theta_0(2^p-1)} = 1.$$

Given $p \in \mathbb{N}$ and asking which values of θ_0 satisfy this equation we get as an answer: $\theta_0(2^p - 1)$ must be an integer n. Asking then for which integers n we get different periodic points $z_n = e^{2\pi i \frac{n}{2^p-1}}$, we see that $n \in \{0, \dots, 2^p - 2\}$ does it. All of the corresponding periodic points lie on the unit circle, and since the points $\frac{n}{2^p-1}$ ($p \in \mathbb{N}$, $0 \le n < 2^p - 1$) form a dense subset of $[0, 1]$ the non-zero periodic points form a dense subset of the unit circle.

What can be said about the behavior of f in the neighbourhood of a periodic point z_0? Suppose its period is p and $z_j := f^{(j)}(z_0)$ $(0 \le j < p)$. Then

$$\begin{aligned}
[f^{(p)}]'(z) &= [f(f^{(p-1)})]'(z) = f'(f^{(p-1)}(z)) \cdot [f^{(p-1)}]'(z) \\
&= \prod_{j=0}^{p-1} f'(f^{(j)}(z)), \\
[f^{(p)}]'(z_0) &= \prod_{j=0}^{p-1} f'(z_j) = [f^{(p)}]'(z_j) \qquad (0 \le j < p) \\
&= \prod_{j=0}^{p-1} (2z_j) = 2^p \prod_{j=0}^{p-1} z_j, \\
\left|[f^{(p)}]'(z_j)\right| &= 2^p > 1.
\end{aligned}$$

The conclusion is that all periodic points with period p (as, in fact, also all other points) on the unit circle are "repellent" with respect to $f^{(p)}$. What, then, happens to the orbits of the various points z? For $|z| = 1$ they are bound to stay and move around on the unit circle. For $|z| < 1$ the orbits converge to the attractive fixed point 0, and for $|z| > 1$ they diverge to ∞. Adjoining the "point" ∞ to $\mathbb{C}$ one obtains a compactification $\mathbb{C}^*$ of $\mathbb{C}$. Since $\mathbb{C}^*$ may also be obtained by projecting every point of $\mathbb{C}$ from the north pole of a sphere, which touches the complex plane tangentially at the south pole, upon the sphere, and adjoining the north pole, $\mathbb{C}^*$ is called the RIEMANN *sphere*. With this understanding the unit circle forms the boundary between the attraction areas – under the iteration of f – of the points 0 and ∞ of the RIEMANN sphere.

This peaceful and beautiful picture is roughly disturbed if instead of $f(z) = z^2$ we consider the function $f(z) = z^2 + c$ for a non-zero complex number c, even if $|c|$ may be arbitrarily small. If $|c|$ is sufficiently small there will still be two fixed points – an "attractive" and a "repellent" one, z_a and z_r, say, – and there will still be a boundary between the points with orbits converging to z_a and the points with orbits "converging" to ∞. This boundary will be invariant under the action of f and will necessarily have to contain all periodic points unequal to z_a. But it seems hard to imagine how this boundary – called a JULIA *set* – might look. This will be what we are after.

3.1 General theory of JULIA sets

In this section we follow the expositions given by [Blanchard, 1984], [Falconer, 1990]. As pointed out before we shall be concerned with the behavior of orbits $\{f^{(k)}(z)\}_{k=1}^{\infty}$ under a polynomial given by

$$f(z) = \sum_{j=0}^{n} c_j z^j \qquad (c_j \in \mathbb{C},\ 0 \leq j \leq n,\ c_n \neq 0) \tag{3.1}$$

of degree $n \geq 2$. Although we eventually shall apply our knowledge to quadratic polynomials only, there is nothing to be gained in the general theory by limiting oneself to $n = 2$, and therefore in this section we shall not do so. Still, we shall require some facts from the theory of functions of a complex variable. To this end, apart from the better known ones, we shall collect some important theorems here in advance. We shall denote the closed disc with radius ρ centered at z by $B(z, \rho)$ and its interior by $U(z, \rho)$. The closure of a set V will be denoted by $\overline{V}$, not to be confused with the complex conjugate $\overline{z}$ of a complex number z.

3.1.1 Theorem (Mapping theorems from the theory of functions of a complex variable).

(a) *(Open mappings) Let f be a holomorphic, non-constant function on a complex domain D (i.e. a connected open subset of $\mathbb{C}$). Then the image $f(E)$ of an open subset E of D is again an open set.* [Cartan, 1966, p. 178]

(b) *(Uniform convergence of holomorphic functions) The limit function f of a sequence $\{f_n\}_{n=1}^{\infty}$ of functions, holomorphic on a complex domain D, which converges uniformly on compact subsets of D is holomorphic in D and the sequence of derivatives $\{f'_n\}_{n=1}^{\infty}$ converges uniformly on compact subsets of D to f'.* [Cartan, 1966, Theorem 1 and 2, p. 145]

(c) *(Inverse function) Suppose the function f is holomorphic in a neighbourhood of $z_0 \in \mathbb{C}$ and suppose $f'(z_0) \neq 0$. Then there exists a neighbourhood U of z_0, a neighbourhood V of the point $w_0 = f(z_0)$, and a function g which is uniquely determined and holomorphic in V, such that $g(w_0) = z_0$ and $w = f(z)$ $(z \in U) \iff z = g(w)$ $(w \in V)$.* [Cartan, 1966, Theorem 1.1, p. 175]

(d) *(Monodromy theorem) The analytic continuation of a function holomorphic in some open subset of a simply connected domain to this simply connected domain is unique.* [Nevanlinna and Paatero, 1965, §2, p. 246]

3.1.2 Definition. A fixed point $z_0 \in \mathbb{C}$ of f $(f(z_0) = z_0)$ is called

$$\begin{aligned} \textit{super-attractive} \quad & \text{if } f'(z_0) = 0, \\ \textit{attractive} \quad & \text{if } |f'(z_0)| < 1, \\ \textit{neutral} \quad & \text{if } |f'(z_0)| = 1, \\ \textit{repellent} \quad & \text{if } |f'(z_0)| > 1. \end{aligned}$$

We shall also consider the point $\infty \in \mathbb{C}^*$ as a fixed point of f. As a consequence of the assertion of Lemma 3.1.10 it should be considered as attractive.

3.1.3 Definition. A point $z_0 \in \mathbb{C}$ is called *p-periodic for f* ($p \in \mathbb{N}$) if $f^{(p)}(z_0) = z_0$. The minimal number $p \in \mathbb{N}$ satisfying $f^{(p)}(z_0) = z_0$ is called the *period* of z_0.

3.1.4 Theorem. *Let z_0 be a p-periodic point for f and let $z_k := f^{(k)}(z_0)$ $(0 \le k < p)$. Then*

$$f^{(p)\prime}(z_k) = \prod_{n=0}^{p-1} f'(z_n).$$

Proof. Without loss of generality we suppose $k = 0$.

$$f^{(p)\prime}(z_0) = f'(f^{(p-1)}(z_0)) \cdot f^{(p-1)\prime}(z_0) = \prod_{k=0}^{p-1} f'(z_k).$$ □

3.1.5 Definition. A periodic orbit $\{z_k = f^{(k)}(z_0)\}_{k=0}^{p-1}$ $(f^{(p)}(z_k) = z_k, 0 \le k \le p-1)$ and each of its points z_k is called

$$\begin{aligned} &\textit{super-attractive} && \text{if } (f^{(p)})'(z_0) = 0, \\ &\textit{attractive} && \text{if } |(f^{(p)})'(z_0)| < 1, \\ &\textit{neutral} && \text{if } |(f^{(p)})'(z_0)| = 1, \\ &\textit{repellent} && \text{if } |(f^{(p)})'(z_0)| > 1. \end{aligned}$$

3.1.6 Definition. The JULIA *set* $J(f)$ is the closure of the set of all repellent periodic points for f. The complement $F(f) = \mathbb{C} \setminus J(f)$ of the JULIA set $J(f)$ is called the FATOU *set* of f.

The content of the rest of this section is to make sure that (just as in the case of $f(z) = z^2$) the JULIA set is not empty, compact and perfect (i.e. it does not contain any isolated points), nowhere dense, invariant under the action of f and $f^{(-1)}$, and that it is the boundary of the set of complex numbers with unbounded orbits. In order to get to know the JULIA set it is advisable to make a detour. The set $J_0(f)$ which we are going to define will turn out to be identical with $J(f)$. But the definition of $J_0(f)$, complicated as it may seem, allows an easier access to the properties of $J_0(f)$.

3.1.7 Definition. A sequence $\{g_n\}_{n=0}^\infty$ of complex-valued functions g_n which are holomorphic in an open set $U \subset \mathbb{C}$ is called *normal in U* if every subsequence of $\{g_n\}_{n=0}^\infty$ again admits a subsequence $\{g_{n_k}\}_{k=0}^\infty$ which converges uniformly on all compact subsets $K \subset U$ – either towards a function holomorphic on U or towards ∞, i.e. which either satisfies

$$\lim_{k,l\to\infty} \sup_{z\in K} |g_{n_k}(z) - g_{n_l}(z)| = 0 \qquad \text{or} \qquad \lim_{k\to\infty} \inf_{z\in K} |g_{n_k}(z)| = \infty.$$

The sequence $\{g_n\}_{n=0}^\infty$ is called *normal in $w \in \mathbb{C}$* if it is normal in some open neighbourhood V of w.

3.1.8 Theorem (MONTEL). *Suppose the sequence $\{g_n\}_{n=0}^{\infty}$ is not normal in an open set $U \subset \mathbb{C}$. Then either $\bigcup_{n=0}^{\infty} g_n(U) = \mathbb{C}$ or there is a $w \in \mathbb{C}$ such that $\bigcup_{n=0}^{\infty} g_n(U) = \mathbb{C} \setminus \{w\}$.*

Proof. See [Saks and Zygmund, 1971, Chapter VII §13, Exercise 2]. □

3.1.9 Definition.

$$\begin{aligned} J_0(f) &:= \{z \in \mathbb{C} : \{f^{(n)}\}_{n=1}^{\infty} \text{ is not normal in } z\}. \\ F_0(f) &:= \{z \in \mathbb{C} : \{f^{(n)}\}_{n=1}^{\infty} \text{ is normal in } z\} = \mathbb{C} \setminus J_0(f). \end{aligned}$$

Evidently the set $F_0(f)$ is open; consequently $J_0(f)$ is closed. We shall simplify the notation by writing $J_0 = J_0(f)$, $F_0 = F_0(f)$.

3.1.10 Lemma. *For f as in* (3.1) *let*

$$r_0(f) := \max\left\{1, \frac{2 + \sum_{j=0}^{n-1} |c_j|}{|c_n|}\right\}.$$

Then for $|z| > r_0(f)$ one has

$$|f(z)| > 2|z|.$$

Consequently

$$\lim_{k \to \infty} f^{(k)}(z) = \infty$$

uniformly on

$$V := \{z : |z| > r_0(f)\} = \mathbb{C} \setminus B(0, r_0(f)). \tag{3.2}$$

Proof. For $|z| > r_0(f)$ we have

$$|c_n|\,|z| - \sum_{j=0}^{n-1} |c_j|\,|z|^{j-n+1} \geq |c_n|\,|z| - \sum_{j=0}^{n-1} |c_j| > 2,$$

$$|f(z)| \geq |z|^{n-1}\Big(|c_n|\,|z| - \sum_{j=0}^{n-1} |c_j|\,|z|^{j-n+1}\Big) > 2|z|. \quad \square$$

3.1.11 Theorem. *J_0 is compact.*

Proof. The sequence $\{f^{(k)}\}_{k=1}^{\infty}$ is normal on the set V as defined in (3.2). Consequently $J_0 \subset B(0, r_0(f))$ is bounded. □

3.1.12 Theorem. $J_0 \neq \emptyset$.

Proof. The fixed points of f are the n solutions z_j $(1 \leq j \leq n)$ of the equation $f(z) = z$. Let $r_1 > \max(r_0(f), |z_1|)$. The open disc $U(0, r_1)$ with center 0 and radius r_1 contains a point z_0 satisfying

$$|f^{(k)}(z_0)| \geq 2^k |z_0| \stackrel{k\to\infty}{\longrightarrow} \infty.$$

Suppose $J_0 = \emptyset$. Then the sequence $\{f^{(k)}\}_{k=1}^{\infty}$ would have to be normal in $U(0, r_1)$. But it is impossible for any subsequence of $\{f^{(k)}\}_{k=1}^{\infty}$ to converge uniformly on a compact subset of $U(0, r_1)$ containing both z_0 and z_1. □

3.1.13 Theorem. $J_0 = f^{-1}(J_0) = f(J_0)$.

Proof. Note that the map $f : \mathbb{C} \to \mathbb{C}$ is surjective. We shall show $F_0 = f^{-1}(F_0) = f(F_0)$. This implies $f^{-1}(J_0) = f^{-1}(\mathbb{C} \setminus F_0) = \mathbb{C} \setminus F_0 = J_0$ and $f(J_0) = J_0$. To simplify the notation again we shall write $\{f^{(n)}\}_n$ for $\{f^{(n)}\}_{n=1}^{\infty}$.

Claim 1: $f^{(-1)}(F_0) \subset F_0$. Since f is surjective this implies $F_0 \subset f(F_0)$.
Proof: Let $z_0 \in f^{(-1)}(F_0)$ be given. We shall exhibit an open neighbourhood U of z_0 on which the sequence $\{f^{(n)}\}_n$ is normal. To this end, let V be an open neighbourhood of $f(z_0) \in F_0$ on which $\{f^{(n)}\}_n$ is normal, and let $U := f^{(-1)}(V)$. Given a subsequence $\{f^{(n')}\}_{n'}$ of $\{f^{(n)}\}_n$ let $\{f^{(n''-1)}\}_{n''}$ be a subsequence of $\{f^{(n'-1)}\}_{n'}$, uniformly converging on compact subsets of V. Let K be a compact subset of U. Then $f(K) \subset V$ is again compact. The sequence $\{f^{(n''-1)}\}_{n''}$ converges uniformly on $f(K)$ and does so by continuity $\{f^{(n'')}\}_{n''}$ on K.

Claim 2: $f(F_0) \subset F_0$. This will imply $F_0 \subset f^{(-1)}(f(F_0)) \subset f^{(-1)}(F_0)$.
Proof: We have to show: Given $z_0 \in F_0$ there is an open neighbourhood U of $f(z_0)$ on which the sequence $\{f^{(n)}\}_n$ is normal. Let V_0 and V_1 be bounded open neighbourhoods of z_0 satisfying $z_0 \in V_1 \subset \overline{V_1} \subset V_0$ on which $\{f^{(n)}\}_n$ is normal. The map f is open (Theorem 3.1.1(a)). Therefore $U := f(V_1)$ is an open neighbourhood of $f(z_0)$. Given a subsequence $\{f^{(n')}\}_{n'}$ of $\{f^{(n)}\}_n$, let $\{f^{(n''+1)}\}_{n''}$ be a subsequence of $\{f^{(n'+1)}\}_{n'}$ which converges uniformly on compact subsets of V_0. Let K be a compact subset of $U = f(V_1)$. Then $K_0 := f^{(-1)}(K) \cap \overline{V_1}$ is a compact subset of V_0. The sequence $\{f^{(n''+1)}\}_{n''}$ converges uniformly on K_0, therefore so does the sequence $\{f^{(n'')}\}_{n''}$ on $f(K_0) = f(f^{(-1)}(K) \cap \overline{V_1}) = K$. □

3.1.14 Theorem. *F_0 contains every attractive periodic orbit of f.*

Proof. Let p be the period of z_0 under f and let $z_j := f^{(j)}(z_0)$ $(0 \leq j < p)$. It suffices to show $z_0 \in F_0$. Choose any $s \in]\,|(f^{(p)})'(z_0)|, 1\,[$ and any $d > \max\{|(f^{(j)})'(z_0)| : 0 \leq j < p\}$. Let δ be so small that for all $z \in V = U(z_0, \delta)$ one has

$$\begin{aligned} |f^{(j)}(z) - z_j| &< d \cdot |z - z_0| \qquad (0 \leq j < p), \\ |f^{(p)}(z) - z_0| &< s \cdot |z - z_0| \; (< \delta) \qquad \text{and therefore} \\ f^{(p)}(z) &\in V. \end{aligned}$$

Let $\{f^{(k')}\}_{k'}$ be a subsequence of $\{f^{(k)}\}_k$. For some j $(0 \leq j < p)$ there has to be a subsequence $\{f^{(k'')}\}_{k''}$ of $\{f^{(k')}\}_{k'}$ satisfying $k'' \equiv j \pmod p$ for all k''. If $z \in V$ then

$$\begin{aligned} |f^{(k'')}(z) - z_j| &= |f^{(mp+j)}(z) - z_j| \\ &= |f^{(j)}(f^{(mp)}(z)) - f^{(j)}(z_0)| \\ &< d \cdot |f^{(mp)}(z) - z_0| \\ &\leq d \cdot s^m \cdot |z - z_0|. \end{aligned}$$

Therefore the sequence $\{f^{(k'')}\}_{k''}$ converges uniformly on V to the constant function z_j. □

3.1.15 Theorem. *J_0 contains every repellent periodic orbit of f. Consequently one has $J(f) \subset J_0$.*

Proof. Let z_0 be a repellent periodic point for f with period p. z_0 is a repellent fixed point of the polynomial $g = f^{(p)}$. Suppose the sequence $\{g^{(k)}\}_k$ is normal in z_0. Then there would be a neighbourhood V of z_0 and a subsequence $\{g^{(k')}\}_{k'}$ of $\{g^{(k)}\}_k$ converging uniformly on every compact subset $K \subset V$. Since z_0 is a fixed point of g, the limit function g_0 has to be finite and holomorphic in V. The same holds for $g_0' = \lim_{k' \to \infty}(g^{(k')})'$ (Theorem 3.1.1(b)). But z_0 is a repellent fixed point of g. Therefore one has $\lim_{k' \to \infty}(g^{(k')})'(z_0) = \lim_{k' \to \infty}(g'(z_0))^{k'} = \infty$. Consequently the sequence $\{g^{(k)}\}_k$ is not normal in z_0 and $z_0 \in J_0$. Finally, since J_0 is closed, we also have $J(f) \subset J_0$. □

3.1.16 Definition. Let z_0 be an attractive fixed point for f (either $z_0 \in \mathbb{C}$ or $z_0 = \infty$). The set $A(f, z_0) := \{z \in \mathbb{C} : \lim_{k \to \infty} f^{(k)}(z) = z_0\}$ is called the *attractive basin* of z_0.

Note that as an immediate consequence of the definition one gets $A(f, z_0) = f^{(-1)}(A(f, z_0))$.

3.1.17 Theorem. *If z_0 is an attractive fixed point for f, then $A(f, z_0)$ is open and $A(f, z_0) \subset F_0$.*

Proof. Let first $z_0 \neq \infty$. Choose $s > 0$ so as to satisfy $|f'(z_0)| < s < 1$. Then for a suitable $\delta > 0$ and the open disc $V = U(z_0, \delta)$ one has

$$z \in V \quad \Longrightarrow \quad \frac{|f(z) - z_0|}{|z - z_0|} < s,$$

and consequently $V \subset A(f, z_0)$. As in the proof of Theorem 3.1.14 the sequence $\{f^{(k)}\}_k$ converges on V uniformly to the constant z_0. Consequently one has $V \subset F_0$. Given $z \in A(f, z_0)$ there is a $k \in \mathbb{N}$ satisfying $f^{(k)}(z) \in V$. We obtain

$$A(f, z_0) = \bigcup_{k=0}^{\infty} f^{(-k)}(V) \subset \bigcup_{k=0}^{\infty} f^{(-k)}(F_0) = F_0.$$

Suppose now $z_0 = \infty$. Choose $V \subset F_0$ as in (3.2). Then, as before, the set

$$A(f, \infty) = \bigcup_{k=0}^{\infty} f^{(-k)}(V) \subset \bigcup_{k=0}^{\infty} f^{(-k)}(F_0) = F_0$$

is open. □

3.1.18 Definition. For $z \in F_0$ let $C(f, z)$ be the connected component of F_0 containing z.

If $\{z_j = f^{(j)}(z_0)\}_{j=0}^{p-1}$ $(f^{(p)}(z_0) = z_0)$ is an attractive periodic orbit, then by Theorem 3.1.17 for $0 \leq j < p$ one has $C(f, z_j) \subset A(f^{(p)}, z_j)$. Otherwise there would be a point $\tilde{z} \in \partial A(f^{(p)}, z_j) \cap C(f, z_j) \subset F_0$; for every neighbourhood $V \subset C(f, z_j)$ of $\tilde{z}$ the sequence $\{f^{(kp)}(z)\}_k$ would converge to z_j on $V \cap A(f^{(p)}, z_j)$ but remain outside of $A(f^{(p)}, z_j)$ on $V \setminus A(f^{(p)}, z_j)$; this contradicts $\tilde{z} \in F_0$. As a consequence, the sets $C(f, z_j) \subset A(f^{(p)}, z_j)$ have to be disjoint.

3.1.19 Theorem. *Let $\{z_j = f^{(j)}(z_0)\}_{j=0}^{p-1}$ $(f^{(p)}(z_0) = z_0 \in \mathbb{C})$ be an attractive periodic orbit for f. Then the set $E = \bigcup_{j=0}^{p-1} C(f, z_j) \subset \bigcup_{j=0}^{p-1} A(f^{(p)}, z_j) \subset F_0$ contains at least one critical value of f, i.e. a number $w = f(z)$, the pre-image z of which satisfies $f'(z) = 0$.*

Proof. Let us assume the converse hypothesis. We shall lead it to a contradiction in a series of steps.

Claim 1: $f'(z) \neq 0$ for all $z \in E$.
Proof: For $0 \leq j < p$ we shall write $j' := j + 1 \pmod p$. By Theorem 3.1.1(a) the set $f(C(f, z_j))$ is open. It contains $z_{j'}$ and, by continuity, it is connected. Therefore one has $f(C(f, z_j)) \subset C(f, z_{j'})$ and $f(E) \subset E \neq \mathbb{C}$. If for some $z \in C(f, z_j)$ we had $f'(z) = 0$, then $f(z) \in C(f, z_{j'})$ would be a critical value of f in E, the possibility of which we have excluded.

Claim 2: The functions $g_k = f^{(-kp)}$ $(1 \leq k < \infty)$ are well defined and form a normal family on a sufficiently small simply connected neighbourhood of z_0.
Proof: Note that, by claim 1, $[f^{(k)}]'(z) = \prod_{j=0}^{k-1} f'(f^{(j)}(z)) \neq 0$ for all $k \in \mathbb{N}$ and all $z \in E$. By Theorem 3.1.1(c) the function $f^{(kp)} = (f^{(p)})^{(k)}$ $(k \in \mathbb{N})$ is locally invertible at each $z \in C(f, z_0)$. Let $U \subset C(f, z_0)$ be a simply connected neighbourhood of z_0. By Theorem 3.1.1(d) the inverse function g_k of $f^{(kp)}$, obtained by starting with $g_k(z_0) = z_0$ and analytic continuation to all of U, is uniquely defined and holomorphic. By MONTEL's Theorem 3.1.8 the sequence of holomorphic functions $\{g_k\}_k$ on U must be normal since $g_k(U) \subset C(f, z_0) \subset E$ for all $k \in \mathbb{N}$.

Claim 3: z_0 is a repellent fixed point for the sequence $\{g_k\}_{k=1}^{\infty} = \{g_1^{(k)}\}_{k=1}^{\infty}$.
Proof: We have $g_k(z_0) = f^{(-kp)}(z_0) = z_0$ and $s = |g_1'(z_0)| = 1/|(f^{(p)})'(z_0)| > 1$.

As in the proof of Theorem 3.1.15, claim 3 implies that the sequence $\{g_k\}_{k=1}^{\infty}$ cannot be normal in z_0, in contradiction with claim 2. □

3.1.20 Theorem. *The number of different attractive periodic orbits is less than n.*

Proof. Suppose $\{z_{1,j}\}_{j=1}^{p_1}$ and $\{z_{2,j}\}_{j=1}^{p_2}$ are two different attractive periodic orbits. Without loss of generality we may suppose that $C(f, z_{1,0})$ and $C(f, z_{2,0})$ contain the critical values w_1 and w_2 respectively of f. If $w_1 = w_2$ then $C(f, z_{1,0}) = C(f, w_1) = C(f, z_{2,0})$. This implies $z_{1,0} = \lim_{k\to\infty} f^{(kp_1)}(z_{2,0}) = \lim_{k\to\infty} f^{(kp_1p_2)}(z_{2,0}) = z_{2,0}$ which is impossible. Therefore different attractive periodic orbits must correspond to different critical values of f. But there are at most $n-1$ of those. □

At this point one wonders about the number N of neutral periodic points. The idea is that maybe a small variation of the function f may induce also a small change of the periodic points in such a way as to keep the attractive periodic points attractive but change at least some of the originally neutral periodic points into attractive ones and then to apply Theorem 3.1.20. In fact it may be shown that changing the function f to f_w given by $f_w(z) = (1-w)f(z) + w$, taking $w \in \mathbb{C}$ sufficiently small and with a suitable argument, this is accomplished with at least $N/2$ originally neutral periodic points changing into attractive ones. So the number of non-repellent periodic points cannot exceed $2n-2$. Unfortunately the argument is too elaborate to be reproduced here. It may be found in [Blanchard, 1984, Theorem 5.12, p. 111]. Furthermore, DOUADY [Douady, 1983] has shown that the total number of non-repelling periodic points is even less than n. So we have to be content here with an assertion to be proved elsewhere:

3.1.21 Theorem. *The number of neutral periodic orbits is finite.*

Proof. See [Blanchard, 1984, p. 111]. □

3.1.22 Theorem. *If the orbits of all critical points of f are unbounded, then there are no attractive periodic orbits of f.*

Proof. Our hypothesis implies that $\lim_{k\to\infty} f^{(k)}(w) = \infty$ for all critical values w of f. Suppose $\{z_j\}_{j=0}^{p-1}$ is an attractive periodic orbit of f. Without loss of generality we suppose that $C(f, z_0)$ contains a critical value w of f. Then $\lim_{k\to\infty} f^{(kp)}(w) = z_0$ in contradiction to our hypothesis. □

3.1.23 Theorem. $J_0(f^{(m)}) = J_0(f)$ *for all* $m \in \mathbb{N}$.

Proof. We shall show $F_0(f^{(m)}) = F_0(f)$.

Claim 1: $F_0(f) \subset F_0(f^{(m)})$.
Proof: Suppose the sequence $\{f^{(k)}\}_k$ is normal on the open neighbourhood U of $z_0 \in F_0$. A subsequence $\{(f^{(m)})^{(k')}\}_{k'} = \{(f^{(m\cdot k')}\}_{k'}$ of the sequence $\{(f^{(m)})^{(k)}\}_k$ is also a subsequence of the sequence $\{f^{(k)}\}_k$. Therefore there is a subsequence $\{f^{(m\cdot k'')}\}_{k''}$ of $\{f^{(m\cdot k')}\}_{k'}$, which converges uniformly on all compact subsets of U.

Claim 2: $F_0(f^{(m)}) \subset F_0(f)$.
Proof: Suppose the sequence $\{(f^{(m)})^{(k)}\}_k$ is normal on the open neighbourhood U of $z_0 \in F_0(f^{(m)})$, and let $\{f^{(k')}\}_{k'}$ be a subsequence of the sequence $\{f^{(k)}\}_k$. At least

for one r $(0 \le r < m)$ there is a subsequence of $\{f^{(k')}\}_{k'}$ of the form $\{f^{(m\cdot k''+r)}\}_{k''}$. The sequence $\{f^{(m\cdot k'')}\}_{k''}$ contains a subsequence $\{f^{(m\cdot k''')}\}_{k'''}$, which converges uniformly on all compact subsets of U. Consequently, also the sequence $\{f^{(m\cdot k'''+r)}\}_{k'''}$ converges uniformly on all compact subsets of U. □

3.1.24 Theorem. *Let f be a polynomial of degree $n > 1$ and let U be a neighbourhood of a point $w \in J_0$. Then*

$$\bigcup_{k=1}^{\infty} f^{(k)}(U) \supset J_0$$

and either

$$\bigcup_{k=1}^{\infty} f^{(k)}(U) = \mathbb{C}$$

or there exists a number $c \in \mathbb{C}$ and a fixed point $z_0 \in F_0(f)$ of f satisfying

$$\bigcup_{k=1}^{\infty} f^{(k)}(U) = \mathbb{C} \setminus \{z_0\} \qquad \text{and} \qquad f(z) = z_0 + c\cdot(z - z_0)^n.$$

Proof. The sequence $\{f^{(k)}\}_{k=1}^{\infty}$ is not normal on U. Suppose $\mathbb{C} \ne W = \bigcup_{k=1}^{\infty} f^{(k)}(U)$. Then, by the MONTEL's Theorem 3.1.8, there exists a point $z_0 \in \mathbb{C}$ satisfying $W = \mathbb{C} \setminus \{z_0\}$. Let z_i $(1 \le i \le n)$ be the roots of the equation $f(z) - z_0 = 0$. If for some i $(1 \le i \le n)$ we had $z_i \in W$, then $z_0 = f(z_i) \in f(W) \subset W$, the possibility of which we have excluded. As a consequence we get $z_i = z_0$ $(1 \le i \le n)$ and $f(z) - z_0 = c\cdot(z - z_0)^n$. Suppose now $|z - z_0| \le r := (2|c|)^{-1/(n-1)}$. We get

$$|f(z) - z_0| = |c|\cdot|z - z_0|^n \le \frac{|c|}{2|c|}\cdot|z - z_0| = \frac{1}{2}|z - z_0|,$$

$$|f^{(k)}(z) - z_0| \le \frac{1}{2^k}|z - z_0| \le \frac{r}{2^k}.$$

Therefore, on the closed disc $B(z_0, r)$ the sequence $f^{(k)}(z)$ converges uniformly to the constant z_0. As a consequence, the sequence $\{f^{(k)}\}_{k=1}^{\infty}$ is normal in z_0. □

3.1.25 Theorem. *Let U be an open set satisfying $U \cap J_0 \ne \emptyset$, and let $w \in \bigcup_{k=1}^{\infty} f^{(k)}(U)$ be given. Then $f^{(-k)}(w) \cap U \ne \emptyset$ for infinitely many $k \in \mathbb{N}$.*

Proof. If $W = \bigcup_{k=1}^{\infty} f^{(k)}(U) \ne \mathbb{C}$ let z_0 be defined as in Theorem 3.1.24. Let $k_1 := \min\{k \in \mathbb{N} : w \in f^{(k)}(U)\}$. There exists a point $w_1 \in U$ such that $f^{(k_1)}(w_1) = w$. In case $W \ne \mathbb{C}$ we have $w_1 \ne z_0$ since $f^{(k_1)}(z_0) = z_0 \ne w$. Consequently, in any case, we have $w_1 \in W$. By induction we may therefore construct a monotone increasing sequence $\{k_j\}_{j=1}^{\infty} \subset \mathbb{N}$ and a sequence $\{w_j\}_{j=1}^{\infty} \subset \mathbb{C}$ satisfying $w_j \in f^{(-k_j)}(w) \cap U$ $(1 \le j < \infty)$. □

3.1.26 Theorem. *For every point $w \in J_0$ one has*

$$J_0 = \overline{\bigcup_{k=1}^{\infty} f^{(-k)}(w)}.$$

Proof. For $w \in J_0$ Theorem 3.1.13 furnishes $f^{(-k)}(w) \subset J_0$ and

$$\overline{\bigcup_{k=1}^{\infty} f^{(-k)}(w)} \subset J_0.$$

Let U be an arbitrary open set meeting J_0. By Theorem 3.1.24 we have

$$w \in \bigcup_{k=1}^{\infty} f^{(k)}(U)$$

and therefore

$$\bigcup_{k=1}^{\infty} f^{(-k)}(w) \cap U \neq \emptyset \qquad \text{and} \qquad \overline{\bigcup_{k=1}^{\infty} f^{(-k)}(w)} \supset J_0.$$ □

3.1.27 Theorem. *J_0 is nowhere dense.*

Proof. If J_0 contained an open set U, Theorem 3.1.13 would yield $\bigcup_{k=1}^{\infty} f^{(k)}(U) \subset J_0$. By Theorem 3.1.24 this is impossible since J_0 is compact (Theorem 3.1.11). □

3.1.28 Theorem. *J_0 is perfect, i.e. does not contain any isolated points.*

Proof. Let U be an open neighbourhood of a point $w_0 \in J_0$. We want to show that U contains a point $w \in J_0$ different from w_0. We distinguish three different cases.

(a) w_0 is not a periodic point of f. Then all pre-images of w_0 under f are mutually different. Let $w_0 = f(w_1)$. Then $w_1 \neq w_0$ and Theorems 3.1.13 and 3.1.24 yield $w_1 \in J_0 \subset \bigcup_{k=1}^{\infty} f^{(k)}(U)$ and $w_0 \notin f^{(-k)}(w_1) \subset J_0$. However, by Theorem 3.1.25, one has $f^{(-k)}(w_1) \cap U \neq \emptyset$ for infinitely many $k \in \mathbb{N}$.

(b) $w_0 = f(w_0)$ is a fixed point of f. If all pre-images $f^{(-1)}(w_0)$ coincide with w_0, then as shown in the proof of Theorem 3.1.24 $f(z) = w_0 + c(z - w_0)^n$ and $w_0 \in F_0$ which is impossible. If $w_1 \neq w_0$ and $f(w_1) = w_0$, then as in case (a) we get $w_0 \notin f^{(-k)}(w_1) \subset J_0$ and $f^{(-k)}(w_1) \cap U \neq \emptyset$ for infinitely many $k \in \mathbb{N}$.

(c) $w_0 = f^{(p)}(w_0)$ $(p > 1)$. The reasoning already employed in case (b) applies for the polynomial $f^{(p)}$ and the set $J_0(f^{(p)}) = J_0(f)$ (Theorem 3.1.23). □

3.1.29 Theorem. $J(f) = J_0$.

Proof. In Theorem 3.1.15 it has already been shown that $J(f) \subset J_0$. So it remains to show $J(f) \supset J_0$.

There are only finitely many fixed points and critical values of f. Eliminating these points from J_0 we obtain a set

$$J_1 = \{w \in J_0 : (f(w) \neq w) \wedge (f'(z) \neq 0 \; \forall z \in f^{(-1)}(w))\}. \tag{3.3}$$

Since J_0 does not contain any isolated points we have $\overline{J_1} = J_0$. We shall substantiate our claim by showing $J_1 \subset J(f) = \overline{J(f)}$.

Given $w_0 \in J_1$, by (3.3) there is a $z_0 \in J_0 \setminus \{w_0\}$ satisfying the relations $f(z_0) = w_0$, $f'(z_0) \neq 0$. Then by Theorem 3.1.1(c) there are two disjoint arbitrarily small open neighbourhoods U of z_0 and W of w_0 such that f maps U bijectively upon W. Denote this bijection by $\tilde{f}$. The union P of all non-repellent periodic orbits is finite by Theorems 3.1.20 and 3.1.21. Therefore the sets U and W may be chosen disjoint from P. Define a sequence of functions h_k which are holomorphic on W by

$$h_k(w) := \frac{f^{(k)}(w) - w}{\tilde{f}^{(-1)}(w) - w} \qquad (w \in W,\ k \in \mathbb{N}).$$

If the sequence $\{h_k\}_{k=1}^{\infty}$ were normal on W, then so would the sequence $\{f^{(k)}\}_{k=1}^{\infty}$ have to be, which is impossible because of $w_0 \in J_0$. By Theorem 3.1.8 of MONTEL the set $\bigcup_{k=1}^{\infty} h_k(W)$ contains at least one of the numbers 0 and 1. Consequently, either there is a $w_1 \in W$ and a $k_1 \in \mathbb{N}$ satisfying $h_{k_1}(w_1) = 0$, or there is a $w_2 \in W$ and a $k_2 \in \mathbb{N}$ satisfying $h_{k_2}(w_2) = 1$. In the first case we obtain $f^{(k_1)}(w_1) = w_1$, in the second case we obtain $f^{(k_2+1)}(w_2) = w_2$. We conclude that the point w_0 may be approximated arbitrarily close by repellent periodic points. It therefore must belong to $J(f)$. □

3.1.30 Theorem. *If z_0 is an attractive fixed point for f or if $z_0 = \infty$, then $\partial A(f, z_0) = J(f)$.*

Proof. In any case we have $z_0 \notin J(f)$.

Claim 1: $J(f) \subset \partial A(f, z_0)$.
Proof: Let $z \in J(f)$ be given. By Theorem 3.1.13 its orbit is contained in $J(f)$ and cannot converge to z_0. We conclude $z \notin A(f, z_0)$. Let U be an arbitrary open neighbourhood of z. By Theorem 3.1.24 we have $A(f, z_0) \cap \bigcup_{k=1}^{\infty} f^{(k)}(U) \neq \emptyset$ and consequently $U \cap f^{(-k)}A(f, z_0) = U \cap A(f, z_0) \neq \emptyset$ for at least one (in fact infinitely many) $k \in \mathbb{N}$. We conclude $z \in \partial A(f, z_0)$.

Claim 2: $J(f) \supset \partial A(f, z_0)$.
Proof: Suppose there were a $z \in \partial A(f, z_0) \setminus J_0$. Then there would also be an open neighbourhood V of z and a subsequence $\{f^{(k')}\}_{k'}$ converging uniformly on every compact subset of V either to $z_0 = \infty$ or to a holomorphic function g. On the open set $V \cap A(f, z_0) \neq \emptyset$ the sequence $\{f^{(k')}\}_{k'}$ certainly converges to z_0. Therefore it has to converge on all of V, including on z, to z_0. This would imply $z \in A(f, z_0)$ in contradiction to our hypothesis. □

Let us try to visualize the different situations that occur corresponding to the number of attractive periodic orbits. If there are none of them apart from the attractive fixed point ∞, then $J(f)$ must consist of a nowhere dense but perfect set of "gaps" in the complex plane. If there is exactly one attractive periodic orbit consisting of a fixed point, then one can visualize $J(f)$ as the boundary between its attractive basin and the attractive basin of ∞. As soon as there are more than one attractive fixed points then $J(f)$ is the boundary of each of the corresponding attractive basins – which is going hard on our intuition. The same situation already occurs if there is at least one attractive periodic orbit with period $p > 1$: Each point of this orbit is a fixed point of $f^{(p)}$, and $J(f)$ and $J(f^{(p)})$ coincide by Theorem 3.1.23.

Everything stated in this section holds – in adapted form – also for rational functions, i.e. the quotient of two polynomials [Blanchard, 1984]. We, however, shall be happily content to apply its assertions to the simplest situation, the case of a polynomial f of second order.

3.2 JULIA sets for quadratic polynomials

Properly speaking, a polynomial of second order would look like the equation $g(w) = c_2w^2 + c_1w + c_0$ ($c_2 \neq 0$). In order to simplify the situation let us try the substitution $w = h(z) = az+b$ with the inverse $z = h^{(-1)}(w) = \frac{w-b}{a}$, which amounts to a similarity transformation in the complex plane. Using the notation $g \circ h(z) = g(h(z))$ we get

$$\begin{aligned} g \circ h(z) &= a^2c_2z^2 + a(2bc_2 + c_1)z + c_2b^2 + c_1b + c_0, \\ h^{(-1)} \circ g \circ h(z) &= ac_2z^2 + (2bc_2 + c_1)z + \frac{c_2b^2 + (c_1 - 1)b + c_0}{a}. \end{aligned}$$

By choosing $a = \frac{1}{c_2}$, $b = -\frac{c_1}{2c_2}$, and $c = -\frac{c_2b^2+(c_1-1)b+c_0}{a}$ we obtain

$$\begin{aligned} ac_2 &= 1, \\ 2bc_2 + c_1 &= 0, \\ f(z) &:= h^{(-1)} \circ g \circ h(z) \\ &= z^2 - c, \qquad (3.4) \\ f^{(k)} &= h^{(-1)} \circ g^{(k)} \circ h \qquad \text{for all } k \in \mathbb{N}. \end{aligned}$$

The form (3.4) for f – which we shall use in the sequel – is handy since it provides immediate information on fixed points, critical values, and points with periodic orbits of period 2. For fixed points we only have to solve the equation

$$z^2 - z - c = 0,$$

which furnishes the following answer:

3.2.1 Theorem. *The fixed points of f as in* (3.4) *are*

$$z_{1,2} = \frac{1 \pm \sqrt{1+4c}}{2}.$$

In order to find out whether a fixed point is attractive we investigate the derivative of f.

$$\begin{aligned} f'(z) &= 2z, \\ |f'(z_{1,2})| &= |1 \pm \sqrt{1+4c}\,|. \end{aligned}$$

By Theorem 3.1.20 there can be at most one attractive fixed point. Suppose we choose the complex square root in such a way that $|f'(z_2)| = |1 + \sqrt{1+4c}| \geq 1$. Let us first locate the values c for which $|1-\sqrt{1+4c}|=1$. To this end we write $f'(z_1)=1-\sqrt{1+4c}=e^{i\varphi}$. We obtain

$$\begin{aligned} 1-e^{i\varphi} &= \sqrt{1+4c}, \\ 1-2e^{i\varphi}+e^{2i\varphi} &= 1+4c, \\ 4c &= e^{2i\varphi}-2e^{i\varphi}. \end{aligned}$$

Elementary geometric arguments show that for $-\pi \le \varphi \le \pi$ the points

$$w(\varphi) = 2e^{i\varphi}-e^{2i\varphi} = 2\cos\varphi-\cos 2\varphi+i(2\sin\varphi-\sin 2\varphi)$$

lie on a cycloid, the path of the point 1 on the circle with center 2 and radius 1 rolling on the circle with center 0 and radius 1. Elementary analysis provides the points $w(0)=1$ and $w(\pm\frac{2\pi}{3})=-\frac{1}{2}\pm i\frac{3\sqrt{3}}{2}$ with tangents parallel to the real axis, and the points $w(\pi)=-3$ and $w(\pm\frac{\pi}{3})=\frac{3}{2}\pm i\frac{\sqrt{3}}{2}$ with tangents parallel to the imaginary axis.

The looked-for values of c are given by $c=-\frac{w(\varphi)}{4}$, evidently the boundary between those values of c for which $|f'(z_1)|=|1-\sqrt{1+4c}|<1$ and those for which $|f'(z_1)|>1$. Since $c=0$ lying inside the cycloid provides for $|f'(z_1)|=0$ the function p given by $p(c):=f'(z_1(c))=1-\sqrt{1+4c}$ maps the interior of the cycloid $C_0=\{\frac{1}{4}(e^{2i\varphi}-2e^{i\varphi}): 0\le\varphi\le 2\pi\}$ onto the interior of the unit circle. Since the points $f'(z_{1,2})=1\pm\sqrt{1+4c}$ lie symmetric with respect to the point 1 on the unit circle, if $z_1\ne 1$ lies in the unit disc, then z_2 has to lie in the exterior of it, and z_2 has to be a repellent fixed point.

3.2.2 Theorem. *The function $f=z^2-c$ admits an attractive fixed point $z_1=\frac{1-\sqrt{1+4c}}{2}$ if and only if c lies inside the cycloid $C_0=\{\frac{1}{4}(e^{2i\varphi}-2e^{i\varphi}): 0\le\varphi\le 2\pi\}$, and a neutral fixed point $z_1=\frac{e^{i\varphi}}{2}$ with $f'(z_1)=e^{i\varphi}$ if and only if $c=\frac{1}{4}(e^{2i\varphi}-2e^{i\varphi})$. In both cases (except for $c=-\frac{1}{4}$, $\varphi=0$) the fixed point $z_2=\frac{1+\sqrt{1+4c}}{2}$ is repellent and belongs to the* JULIA *set $J(f)$.*

Points z on a periodic orbit of period 2 have to satisfy the equation

$$f^{(2)}(z)-z = 0, \tag{3.5}$$

where

$$\begin{aligned} f^{(2)}(z) &= f(z^2-c) = (z^2-c)^2-c = z^4-2cz^2+c^2-c, \\ f^{(2)}(z)-z &= z^4-2cz^2-z+c^2-c. \end{aligned} \tag{3.6}$$

Since also fixed points have to satisfy equation (3.5), the polynomial (3.6) has to be dividable by the polynomial z^2-z-c, the roots of which are the fixed points. Indeed one has

$$z^4-2cz^2-z+c^2-c = (z^2-z-c)(z^2+z+1-c).$$

There evidently exists precisely one periodic orbit with period 2. Its points are the roots $z_{3,4}$ of the polynomial $z^2+z+1-c$. Whether it is attractive or not is decided by the value

$$|[f^{(2)}]'(z_{3,4})| = |f'(z_3)|\,|f'(z_4)| = 4|z_3z_4| = 4|1-c|.$$

3.2.3 Theorem. *For the polynomial $f(z) = z^2 - c$ there is precisely one periodic orbit $\{z_3, z_4\}$ with period 2, and $z_{3,4} = \frac{1}{2}(-1 \pm \sqrt{4c-3})$. This orbit is attractive if and only if $|1-c| < \frac{1}{4}$.*

The last inequality is satisfied precisely if the point c lies in the interior of a circle with center 1 and radius $\frac{1}{4}$ touching the cycloid C_0 in the point $\frac{3}{4}$.

There is only one critical point $z_0 = 0$ and one critical value $-c$. According to Theorem 3.1.22 it is useless to look for attractive periodic orbits if the orbit of 0 is unbounded. We can be more precise about that. For $c = 2$ we still get $f(2) = 2$, so 2 is a fixed point for f and $f^{(k)}(0) = 2$ for all $k \geq 2$. But a slightly larger value of $|c|$ changes the situation.

3.2.4 Theorem. *If $|z| = 2 + \varepsilon > 2$ and $|z| \geq |c|$, then*

$$|f(z)| \geq (1+\varepsilon)\,|z| \qquad \textit{and} \qquad \lim_{k\to\infty} f^{(k)}(z) = \infty.$$

Consequently, for $|c| > 2$ the sequence $\{f^{(k)}(0)\}_{k=0}^{\infty} = \infty$ is unbounded.

Proof. For $|z| \geq |c|$ we have

$$\begin{aligned} |f(z)| &= \left|z\left(z - \frac{c}{z}\right)\right| = |z|\left|z - \frac{c}{z}\right| \geq |z|\left(|z| - \frac{|c|}{|z|}\right) \\ &\geq |z|\,(2+\varepsilon-1), \\ |f^{(k)}(z)| &\geq (1+\varepsilon)^k\,|z| \stackrel{k\to\infty}{\longrightarrow} \infty, \\ |f^{(k)}(0)| &= |f^{(k-1)}(-c)| \\ &\geq (1+\varepsilon)^{k-1}(2+\varepsilon) \qquad (\text{if } |c| \geq 2+\varepsilon) \\ &\stackrel{k\to\infty}{\longrightarrow} \infty. \end{aligned}$$ □

While $\inf\{\mathbf{Re}\,(c) > 0 : \lim_{k\to\infty} f^{(k)}(0) = \infty\} = 2$ we can get an absolutely smaller lower bound for the set $\{\mathbf{Re}\,(c) < 0 : \lim_{k\to\infty} f^{(k)}(0) = \infty\}$.

3.2.5 Theorem. *If $\mathbf{Re}\,(c) < -1$, then $\lim_{k\to\infty} f^{(k)}(0) = \infty$.*

Proof. Suppose $\mathbf{Re}\,(c) < -1$ and $|c| \leq 2$. We have

$$\begin{aligned} |f^{(2)}(0)| &= |f(-c)| = |c^2 - c| = |c|\,|c-1| > 1 \cdot 2, \\ |f^{(2)}(0)| &= 2+\varepsilon \qquad (\varepsilon > 0), \\ |f^{(k)}(0)| &= |f^{(k-2)}(f^{(2)}(0))| \geq (1+\varepsilon)^{(k-2)}(2+\varepsilon) \stackrel{k\to\infty}{\longrightarrow} \infty. \end{aligned}$$ □

Theorems 3.2.4 and 3.2.5 furnish bounds for the famous set of all c with bounded orbits $\{f^{(k)}(0)\}_{k=1}^{\infty}$ [Mandelbrot, 1980].

3.2.6 Definition. The set M of all parameter values c for which the orbit $\{f^{(k)}(0)\}_{k=1}^{\infty}$ is bounded is called the MANDELBROT *set*.

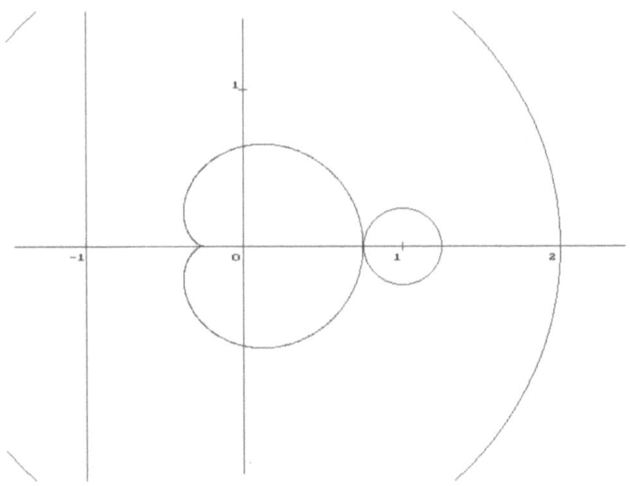

Figure 3.1. The bounds for the MANDELBROT set established in Theorems 3.2.2–3.2.5.

3.3 The MANDELBROT set

The MANDELBROT set serves as a lexicon for the JULIA sets. As already to be expected in view of its definition and of Theorems 3.2.2 and 3.2.3, the location of the parameter c within the MANDELBROT set furnishes information on properties of the corresponding JULIA set. One reason for the importance of this set lies in the assertion of Theorem 3.3.2, for the proof of which we need a lemma. In order to comprehend and visualize its statement it may help to imagine a radar located in the point $-c$, scanning the set G, and a second one, located in the point 0, moved by the apparatus $f^{(-1)}$, turning with half the speed and recording the radial data of the first radar in coded form by taking square roots.

3.3.1 Lemma. *Suppose a simply connected domain G has a smooth boundary ∂G. Both if $-c \in G$ and if $-c \in \partial G$, the set $f^{(-1)}G$ has a smooth boundary and may be represented as the union of two sets $G_0 = \{w = \sqrt{r}\, e^{i\varphi/2} : r \neq 0,\ 0 \le \varphi < 2\pi,\ re^{i\varphi} - c \in G\}$ and $G_1 = \{w = -\sqrt{r}\, e^{i\varphi/2} : r \neq 0,\ 0 \le \varphi < 2\pi,\ re^{i\varphi} - c \in G\} = -G_0$. The sets G_0 and G_1 are central symmetric to each other and f maps each of them one-to-one onto G. If $-c \in G$, then the closure of $f^{(-1)}(G)$ is simply connected. If $-c \in \partial G$, then G_0 and G_1 are open and disjoint, and the boundary $\partial G_0 \cup \partial G_1$ of $f^{(-1)}(G)$ contains precisely one double point at $z = 0$.*

Proof. In both cases considered the boundary ∂G may be written as a differentiable curve $\{z(t) = r(t)e^{i\varphi(t)} - c : 0 \le t \le 1\}$ where $\varphi : [0,1] \to [0, 2\pi]$ is a differentiable function and $z(0) = z(1)$. In the first case ($-c \in G$) a sufficiently small disc $U(-c, \varepsilon)$ is contained in G; we have $r(t) \ge \varepsilon$ for all $t \in [0,1]$ and without loss of generality $\varphi(0) = 0$, $\varphi(1) = 2\pi$. If $-c \in \partial G$ then $z(t)$ starts and ends at $-c = z(0) = z(1)$ and φ is measured counterclockwise starting from the tangent to ∂G in $-c$. Without loss of generality we may suppose $\varphi(0) = 0$, $\varphi(\frac{1}{2}) = \pi$ and $r(0) = r(\frac{1}{2}) = 0$.

In the first case the set G_0 contains inverse images of all points of G, but only half of all inverse images, namely those with argument $\frac{\varphi}{2} \in [0, \pi[$. The set G_1 again contains inverse images of all points of G, to wit the remaining half of the inverse images, namely those with argument $\frac{\varphi}{2}$ in $[\pi, 2\pi[$. The inverse image of the boundary ∂G is the connected union of the two arcs $\{\sqrt{r(t)}\, e^{i\varphi(t)/2} : 0 \le t < 1\}$ obtained by passing once through ∂G, and $\{-\sqrt{r(t)}\, e^{i\varphi(t)/2} = \sqrt{r(t)}\, e^{i(\varphi(t)/2+\pi)} : 0 \le t < 1\}$ obtained by passing once more through ∂G. As a closed continuous curve without multiple points it is the boundary of a simply connected domain consisting of the union of the sets G_0 and G_1.

In the second case the closed continuous curve $\partial G_0 = \{\sqrt{r(t)}\, e^{i\varphi(t)/2} : 0 \le t \le 1\}$ starts at 0 with the real axis as a tangent and passes again (without loss of generality for $t = \frac{1}{2}$ and $\varphi(\frac{1}{2}) = \pi$) through 0 with the imaginary axis as a tangent. The set $G_1 = -G_0$ is bounded by the symmetric image of ∂G_0 with respect to the origin, which continues with the same aforementioned tangent at 0 and ends tangentially on the real axis in 0. Since ∂G does not contain any multiple point, this also holds for its one-to-one pre-images ∂G_0 and ∂G_1. Suppose ∂G_0 and ∂G_1 would intersect in a point $p \ne 0$. Then $p = z(t_1) = -z(t_2)$ where $z(t_1) \in \partial G_0$, $z(t_2) \in \partial G_0$, and $t_1 \ne t_2$; the two different points $z(t_1)$ and $z(t_2)$ on ∂G_0 would have the same image on ∂G under f which is impossible. Consequently, the two domains G_0 and G_1, bounded by ∂G_0 and ∂G_1 respectively, are disjoint. □

3.3.2 Theorem. *The* JULIA *set for the function $f(z) = z^2 - c$ is connected if and only if $c \in M$.*

Proof. Claim 1: If $\{f^{(k)}(0)\}_{k=1}^{\infty}$ is bounded, then $J(f)$ is connected.
Proof: Let $D = B(0, 2)$ be the closed disc with center 0 and radius 2. By Theorem 3.2.4 its complement V_0 in $\mathbb{C}$ is part of the attractive basin $A(f, \infty)$ of ∞ and satisfies $f(V_0) \subset V_0$. This implies $V_0 \subset f^{(-1)}(V_0)$ and $f^{(-1)}(D) \subset D$. Let $V_k := \mathbb{C} \setminus f^{(-k)}(D) = f^{(-k)}(V_0)$. Then $V_k \subset V_{k+1}$, and $A(f, \infty) = \bigcup_{k=1}^{\infty} V_k$. Consequently we have $-c \in K := \bigcap_{k=1}^{\infty} f^{(-k)}(D) = \mathbb{C} \setminus A(f, \infty)$. By Lemma 3.3.1 the sets $f^{(-k)}(D)$ (which are the closures of the open sets $f^{(-k)}U(0, 2)$) are simply connected and so is K. Therefore its boundary $\partial K = \partial A(f, \infty) = J(f)$ (Theorem 3.1.30) is connected.

Claim 2: If $\{f^{(k)}(0)\}_{k=1}^{\infty}$ is unbounded, then $J(f)$ is disconnected.
Proof: Let $D = U(0, r)$ be an open disc with center 0 and radius $r > 2$ such that $J(f) \subset D$ and for some $p \in \mathbb{N}$ we have $f^{(p)}(0) \in \partial D$ and $f^{(k)}(0) \in D$ for $1 \le k < p$. Then $-c = f(0) \in \partial f^{-(p-1)}(D)$. By Lemma 3.3.1 $f^{(-p)}(D)$ is disconnected and consists of two open components D_0 and D_1, each of which is mapped onto $f^{-(p-1)}(D)$. The inclusion $J(f) \subset f^{-(p-1)}(D)$ implies that $J(F) \cap D_1 \ne \emptyset$ and $J(F) \cap D_2 \ne \emptyset$. Consequently $J(f)$ is disconnected. □

3.3.3 Theorem. *The* MANDELBROT *set is closed.*

Proof. Let $c_0 \in \partial M$ be given. Then $|c_0| \le 2$ by Theorem 3.2.4. The orbit of 0 under $f = f_{c_0}$ may be approximated point-wise arbitrarily close by orbits of 0 under functions $f = f_c$ for $c \in M$. These orbits by Theorem 3.2.4 must be contained in $B(0, 2)$. □

3.3.4 The escape time approximation for the MANDELBROT set

With the information acquired so far about bounds and structure of the MANDELBROT set M, one would like to be able to draw an at least approximative picture of M. There is a convenient way to do so with the help of a computer. Establishing the real and imaginary axes with the corresponding unit distances on the screen, we associate with every pixel a complex number z. In particular we want to find out whether, given a $c \in \mathbb{C}$, the orbit of 0 under the corresponding function (3.4) diverges. This happens if and only if for some $k \in \mathbb{N}$ we get $|f^{(k)}(0)| > 2$. So we decide to wait at most until k reaches a certain number n (e.g. $n = 100$ or $n = 500$) and to leave the pixel corresponding to c white if $|f^{(k)}(0)| > 2$ for some $k \leq n$ (the corresponding c certainly does not belong to M) or to paint it e.g. blue if this has not happened so far (the corresponding c is under suspicion of belonging to M) (Figure 3.2). In the resulting blueprint of M we may have enriched M with some points c not belonging to M, since the orbit of 0 under the iterates of the corresponding functions f_c diverges too slowly and its "escape time" exceeds n, but we can adjust the time limit n if we are not satisfied with the resulting picture. Sometimes this has a surprising effect: Where there seemed to be an open subset of M with a strange looking boundary, with an increase of the allowed escape time limit n this subset all but disappears, leaving barely visible traces. These traces, however, signal the existence of more points of M, since it has been shown that M is connected [Douady and Hubbard, 1982]. The reader is kindly asked to bear with the computer which – in several of the following figures – has difficulty catching pixels of all connecting threads belonging to M.

Figure 3.2. An approximating set for the MANDELBROT set: all points $c \in \mathbb{C}$ for which $|f^{(k)}(0)| \leq r = 2$ for all $k \leq n = 100$. The grid may help to locate points c with respect to the MANDELBROT set.

There is an additional feature we can incorporate in this procedure: having e.g. 16 different colours at our disposal we can observe to which residue class $\tilde{k}$ modulo 16 the first exit index k belongs for which $|f^{(k)}(0)| > 2$. If we paint the corresponding pixel with colour number $\tilde{k}$, we partition the complement of M into layers indicating how "close" to M the corresponding values of c lie (the closer to M, the larger the escape time k). In the following figures featuring escape time colour charts these colours appear in the following order: blue (1), green (2), cyan (3), red (4), magenta (5), brown (6), light grey (7), dark grey (8), light blue (9), light green (10), light cyan (11), light red (12), light magenta (13), yellow (14), white (15), black (16≡0). So the blue area in Figure 3.3 indicates that for these values of c one already has $|f(0)| = |c| > 2$, while for the values of c in the neighbouring green area the orbit $\{f^{(k)}(0)\}_{k=0}^{\infty}$ is leaving the disc $B(0, 2)$ already for $k = 2$, and so on. The inner black area (not followed by blue) indicates approximately the proper MANDELBROT set, namely the set of points for which the partial orbit $\{f^{(k)}(0)\}_{k=0}^{100}$ still lies in the disc $B(0, 2)$.

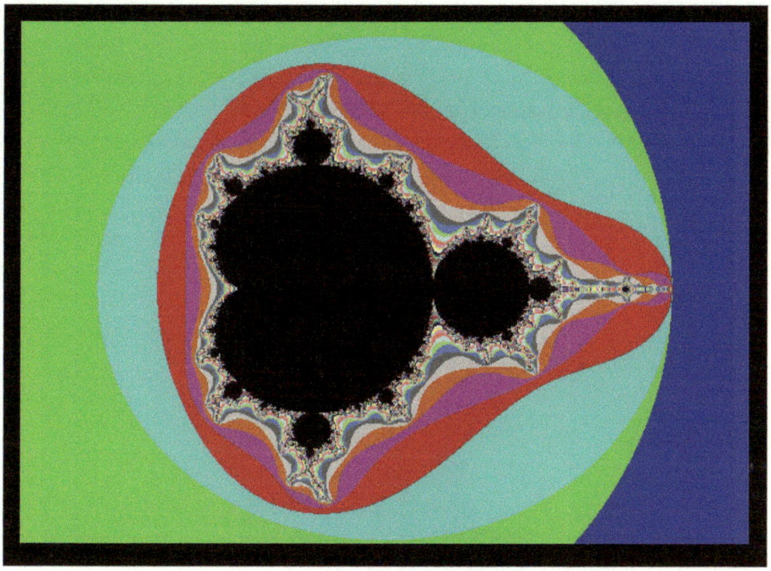

Figure 3.3. An escape time colour chart for the MANDELBROT set M as described in the text.

3.3.5 The escape time approximation for JULIA sets

Essentially the same procedure may be used for an approximate representation of a JULIA set J. In contrast to the former situation, however, now c and with it the function $f = f_c$ is fixed and we test every available z as to whether its orbit diverges. Since $J = \partial A(f, \infty)$ (Theorem 3.1.30), the points of the invariant and compact set J are characterized by the fact that there are neighbouring points whose orbits escape to ∞. According to Lemma 3.1.10 this will certainly be the case as soon as the orbit leaves the closed disc with center 0 and radius $2+|c|$, so this value may serve as a safe escape time bound in place of 2 for the MANDELBROT set. There is, however, a practical difficulty: If c belongs to the MANDELBROT set, then J is the connected boundary of $A(f, \infty)$

(Theorems 3.1.30, 3.3.2), and the pixels corresponding to all points z inside of J are marked blue by our procedure. So we get the set J only as boundary of the complement $\tilde{J}$ of $A(f,\infty)$ which is painted blue (and is sometimes also called the *"filled* JULIA *set"*). The situation is even less satisfying if c does not belong to the MANDELBROT set: As we know, the set J is nowhere dense (Theorem 3.1.27); as a consequence, the chance for a pixel to represent a point of J is small, and since now all neighbouring points of J have orbits diverging to ∞, the pixels of points not extraordinarily close to J will be marked white and a good part of J will remain invisible. A colour chart for the escape time as above may help a little bit in this case.

As an example Figure 3.4 shows an approximation of the JULIA set J for $c = -0.3 + 0.3\,i$ (sorry, not the blue area but only its boundary; c is marked by a grey cross). J is connected and contains the repellent fixed point $z_2 \approx 0.,8565 + 0.4208\,i$ (marked by a red cross) and the repellent periodic orbit consisting of the two points $z_3 \approx -0.3551 + 1.0349\,i$, $z_4 \approx -0.6449 - 1.0349\,i$ (marked by green crosses). There is an attractive fixed point $z_1 \approx 0.1435 - 0.4208\,i$ with $|f'(z_1)| \approx 0.9$. The blue area approximates the attractive basin $A(f, z_1)$. The series of cyan crosses mark the orbit of a point inside of J $(0.87 + 0.43\,i)$ but close to z_2 which of course converges to z_1. Note the five channels along which this convergence takes place; we shall come back to this feature in Section 3.3.7.

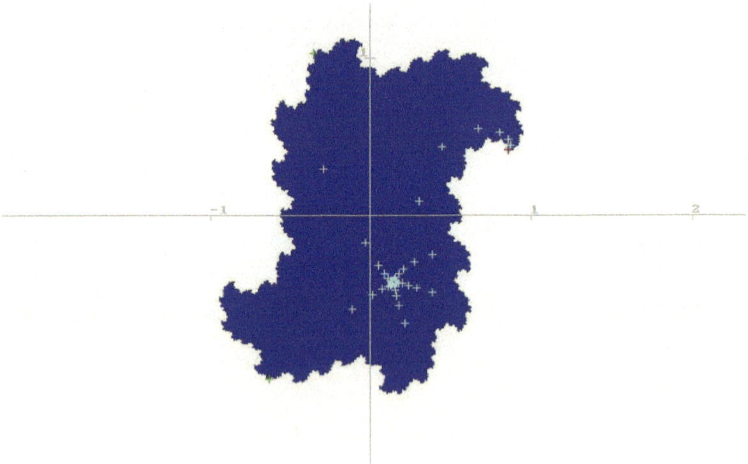

Figure 3.4. The filled JULIA set for $c = -0.3 + 0.3\,i$. Marked by cyan crosses the orbit – converging to the attractive fixed point $z_1 \approx 0.1435 - 0.4208\,i$ – of a point inside of J $(0.87 + 0.43\,i)$ and close to the repellent fixed point $z_2 \approx 0.8565 + 0.4208\,i$.

Figure 3.5 displays in case $c = -0.4 + 0.3\,i$ (marked by a grey cross) what the computer is able to catch of the now disconnected JULIA set J: evidently only some dust marks. Two red crosses mark two repellent fixed points $z_1 \approx 0.1955 - 0.4927\,i$, $z_2 \approx 0.8045 + 0.4927\,i$, two green crosses mark a repellent periodic orbit of period 2 consisting of the points $z_3 \approx -0.3613 + 1.0813\,i$, $z_4 \approx -0.6387 - 1.0813\,i$. The pixel representing (only approximately) the fixed point z_1 has been used for the initial point

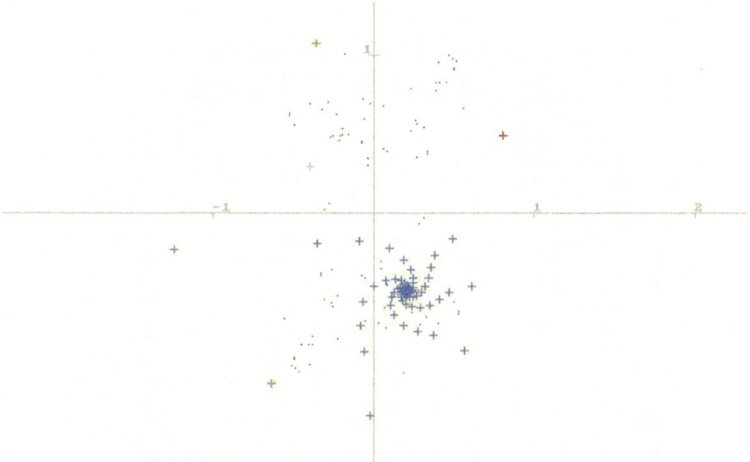

Figure 3.5. The sparse blue points represent pixels which succeed in catching points of the JULIA set for $c = -0.4 + 0.3\,i$. The light blue crosses mark the first 160 points of the orbit of a point close to the repellent fixed point $z_1 \approx 0.1955 - 0.4927\,i$.

of an orbit which now diverges to ∞. The five channels harboring this orbit furnish a striking counterpart to the picture in Figure 3.4 where the attractive fixed point has occupied approximately the same position as here the repellent fixed point z_1.

Figure 3.6 displays the escape time colour chart for the present JULIA set J – but where is J to be found? The answer is: hardly in a recognizable set of pixels; the nowhere dense set J escapes the too coarse grid of pixels. But its presence is indicated in the accumulating sequence of consecutive colours indicating higher and higher escape times, as in the neighbourhood of the point z_1. There one even recognizes the five channels mentioned above. Although now J is (even completely) disconnected and forms the boundary of the attractive basin $A(f, \infty)$ only as a compact and topologically perfect set of gaps in $\mathbb{C}$, there seems to be a conspicuous relationship to the filled JULIA set in Figure 3.4, evidently caused by the closeness of the two corresponding values of c on both sides of the boundary of M.

At this point one starts to wonder how the picture can develop from Figure 3.4 to Figures 3.5 and 3.6, in particular, what happens if the parameter c, wandering from $c_1 = -0.3 + 0.3\,i$ to $c_2 = -0.4 + 0.3\,i$, crosses the boundary of the MANDELBROT set. Traveling on the straight line connecting these two points there certainly must happen something at its intersection $c \approx -0.364852 + 0.3\,i$ with the cycloid $z = \frac{1}{4}(e^{2i\varphi} - 2e^{i\varphi})$ (for $\varphi \approx -69.0512°$, Figure 3.7).

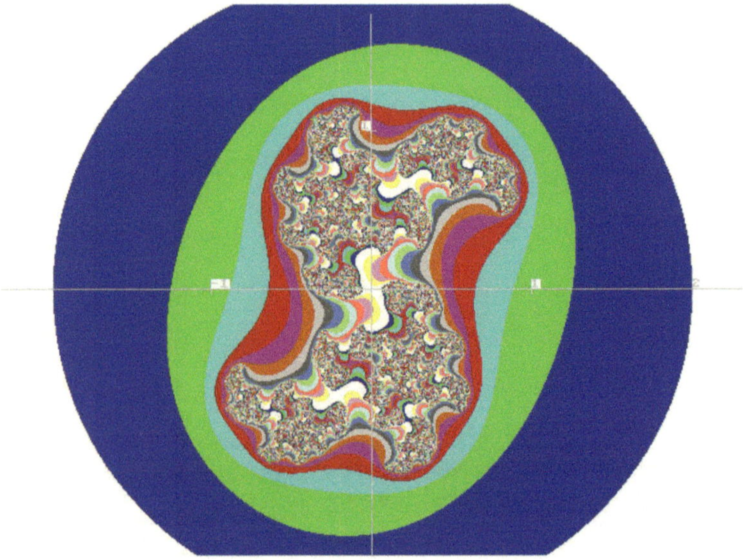

Figure 3.6. An escape time colour chart for the JULIA set in Figure 3.5, corresponding to $c = -0.4 + 0.3\,i$.

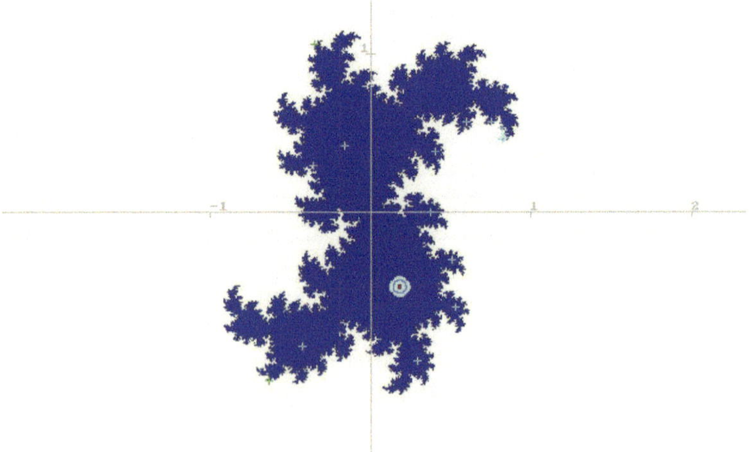

Figure 3.7. The filled JULIA set for the parameter $c \approx -0.3649 + 0.3\,i$ lying on the cycloid C_0 (Theorem 3.2.2) for $\varphi \approx 69.0512°$. The cyan crosses mark the orbit of a point $(0.85 + 0.48\,i)$ close to the repellent fixed point $z_2 \approx 0.8212 + 0.4670\,i$, visiting consecutively and clockwise limbs, tail and trunk of the "dragon" and finally rotating around the neutral fixed point $z_1 \approx 0.1788 - 0.4670\,i$.

Evidently the corresponding filled JULIA set $\tilde{J}$ is on the point of breaking up into pieces, and surprisingly enough the pieces display our well-known feature of self-similarity all over; in fact, we seem to encounter a relative of the dragon in Figure 1.42. According to Theorem 3.2.2 there is a neutral fixed point $z_1 \approx 0.1788 - 0.4670\,i$ in which $f'(z_1) \approx e^{-69.0512\,i}$, and another one $z_2 \approx 0.8212 + 0.4670\,i$ with $|f'(z_2)| \approx 1.8894$. The periodic orbit of period 2 consists of the two points $z_3 \approx -0.3592 + 1.0652\,i)$, $z_4 \approx -0.6408 - 1.0652\,i$. If we let the computer start an orbit $\{f^{(k)}(z_0)\}_{k=0}^{\infty}$ (marked by light blue crosses) at the pixel representing z_2 (it is unable to catch the precise fixed point and belongs to some neighbouring point z_0 at the tip of the tongue of the dragon), then after some first steps, which hardly amount to any visible movement, the points $f^{(k)}(z_0)$ jump to the side of the tongue ($k = 11$) and to the lower lip ($k = 12$), from there to the right front leg on the trunk ($k = 13$), then consecutively to the four right legs connected to the belly ($14 \le k \le 17$), to the tail ($k = 18$), and to the heart in the trunk ($k = 19$). Each of these points assumes in its corresponding part of $\tilde{J}$ roughly the position which z_1 assumes in the main lower body. From there on the following 10000 points of the orbit circle the fixed point z_1 roughly in vertices of pentagons (rotating about angles $\approx 69°$), filling up a closed curve about z_1 without coming close to z_1 – in Figure 3.7 the center line of the cyan belt produced by the cross-marks.

If we check this unexpected behavior by observing the orbit $\{f^{(k)}(0)\}_{k=0}^{10200}$ (Figure 3.8), the computer produces, after the first 10000 points, a string of points – for the eye a closed curve, again the center line in the cyan belt of cross-marks following the outskirts of the lower main body of the filled JULIA set – without caring for the encircled fixed point z_1 or even the smaller annulus in Figure 3.7. The action of f in this region seems to be topologically equivalent to a rotation (cf. [Blanchard, 1984, §7]).

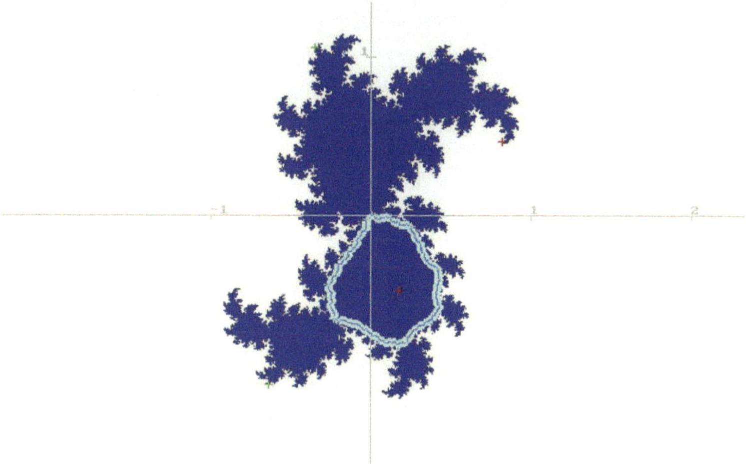

Figure 3.8. The filled JULIA set as in Figure 3.7 with the orbit of the critical point 0, the blue line in the center of the cyan belt.

The special features encountered in the last studied JULIA set suggest that the boundary of the MANDELBROT set deserves special interest. We will focus on this topic in the next section.

3.3.6 Zooming into ∂M

As soon as we magnify parts of the MANDELBROT set we encounter the practical difficulty mentioned above: we want to picture the set $M = \{c \in \mathbb{C} : |f_c^{(k)}(0)| \leq 2 \ \forall k \in \mathbb{N}\}$, but not being able to wait infinitely long we must content ourselves with a picture of the larger approximating set $M(r, n) := \{c \in \mathbb{C} : |f_c^{(k)}(0)| \leq r \ \forall k \leq n\} \supset M$. We still have the choice of r and n, and as $r \geq 2$ gets smaller and n gets larger, the set $M(r, n)$ shrinks towards M. Without trouble we can always take $r = 2$, but magnifying smaller and smaller cuts at the boundary of M we meet more and more points close to ∂M with large escape time. If we want to reasonably distinguish details we are bound to increase n (and thereby the effort and time of computing). Still, if we simply want to enjoy aesthetically whatever pictures are provided for us by the dynamics of the mapping f_c, it seems perfectly legitimate also to care for pictures of sets $M(r, n)$ which avoid fuzzing too much with details. (The size of the figures in this section is, in fact, a little bit smaller than indicated since a margin originally carrying some data information has been eliminated.)

Let us have a closer look at the cut indicated by a red rectangle in Figure 3.9 and depicted magnified in Figure 3.10. There appear armies of buds magnified again in Figure 3.11 which ask for to be looked at closer distance.

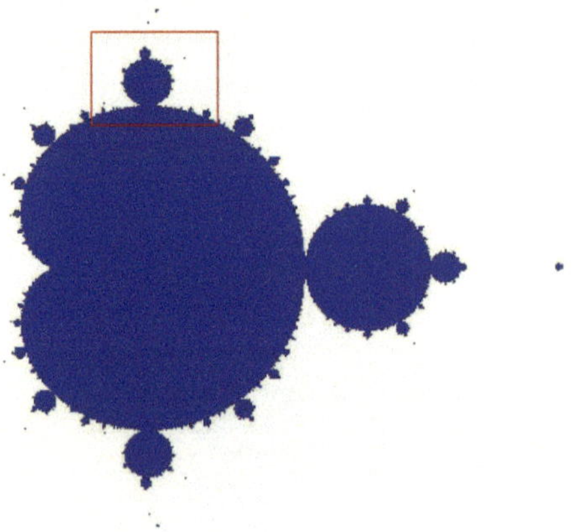

Figure 3.9. The red rectangle marks the area shown in Figure 3.10.

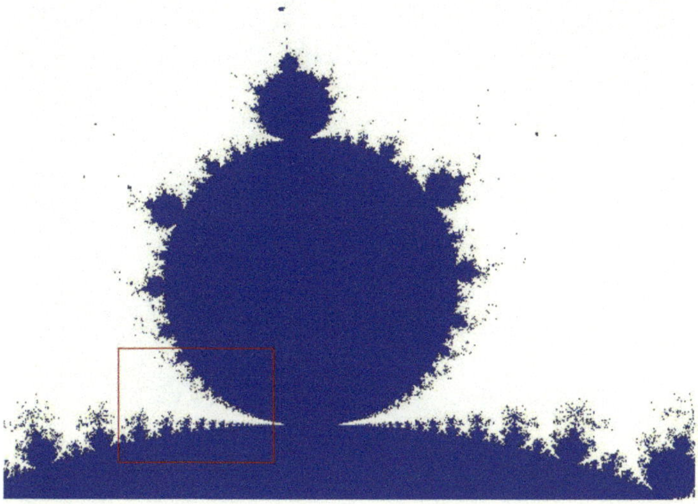

Figure 3.10. $-0.1 \leq x \leq 0.4$, $0.575 \leq y \leq 0.95$; $r = 5$, $n = 100$. The red rectangle marks the area shown in Figure 3.11.

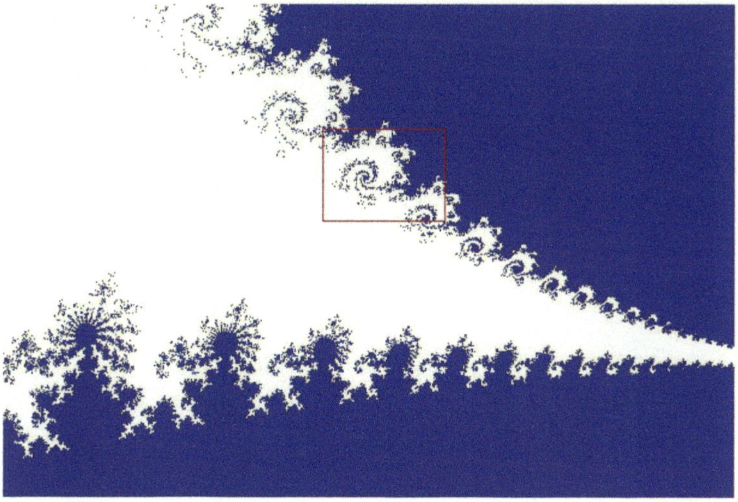

Figure 3.11. $0 \leq x \leq 0.1$, $0.625 \leq y \leq 0.7$; $r = 10$, $n = 100$. The red rectangle marks the area shown in Figure 3.12.

Figure 3.12. $0.044 \leq x \leq 0.06$, $0.669 \leq y \leq 0.681$; $r = 10$, $n = 100$.

Figure 3.13. Same figure as Figure 3.11, but now the red rectangle marks the area shown in Figure 3.14.

Figure 3.14. $0.08 \leq x \leq 0.084,\ 0.648 \leq y \leq 0.651;\ r = 10,\ n = 200.$

The "hanging" ones already appear rather fantastic (Figure 3.12), but how about the "standing" ones indicated again by a red rectangle in Figure 3.13? With your indulgence, my first impression of Figure 3.14 was: three cows are curious about the spiral above their heads (we shall meet other members of this population in later figures again). Here the spell is lifted as soon as we increase the escape time bound from 200 to 500 as done in Figure 3.15: the heads are dissolved, together with the other large blue areas of Figure 3.14, into radiating flowers, the centers of which would still dissolve further if we would further increase the escape time bound. Still, the structure of the spiral seems to be interesting. Magnifying the red rectangle again produces a strip resembling a crafty bracelet (the blue discs standing for gems, Figure 3.16). The middle piece magnified again (Figure 3.17) furnishes an area suspiciously looking like the original MANDELBROT set, but now in a luxurious decoration (Figure 3.18) in which spirals and rosettes abound (Figure 3.19). We stop zooming here since we have to stop somewhere, but there seems to be no end to beauty in the abysses of this microcosm. (This series of pictures has been inspired by similar ones in [Peitgen and Richter, 1986, maps 34–50] and [Peitgen, Jürgens, Saupe, 1992, 14.3].)

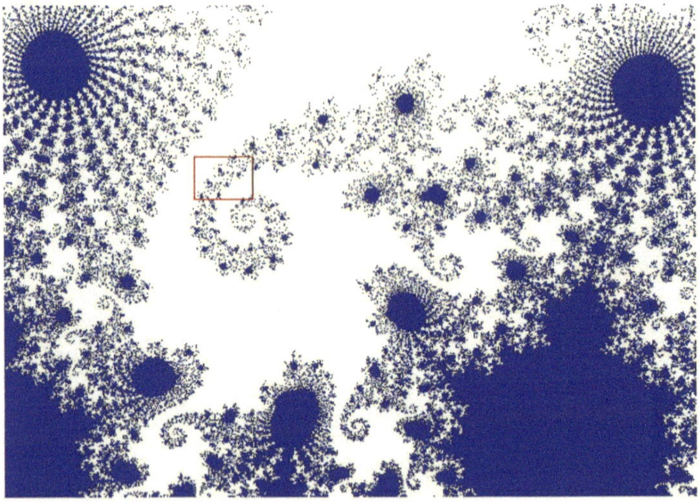

Figure 3.15. Same rectangle as in Figure 3.14, but $r = 10$, $n = 500$. The red rectangle marks the area shown in Figure 3.16.

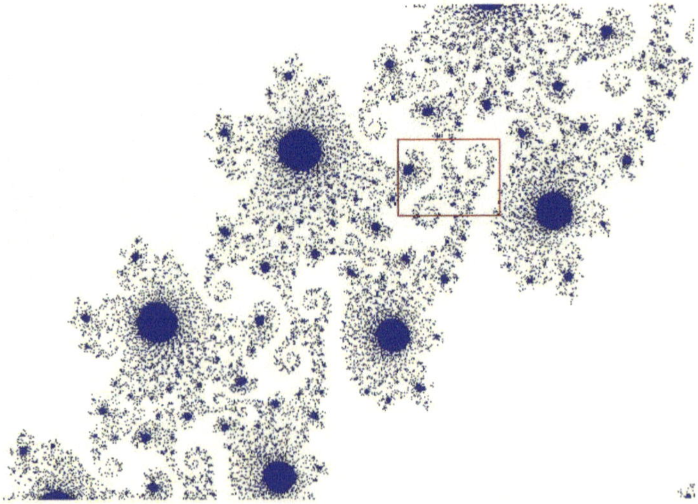

Figure 3.16. $0.0812 \leq x \leq 0.0815$, $0.649775 \leq y \leq 0.65$; $r = 2$, $n = 500$. The red rectangle marks the area shown in Figure 3.17.

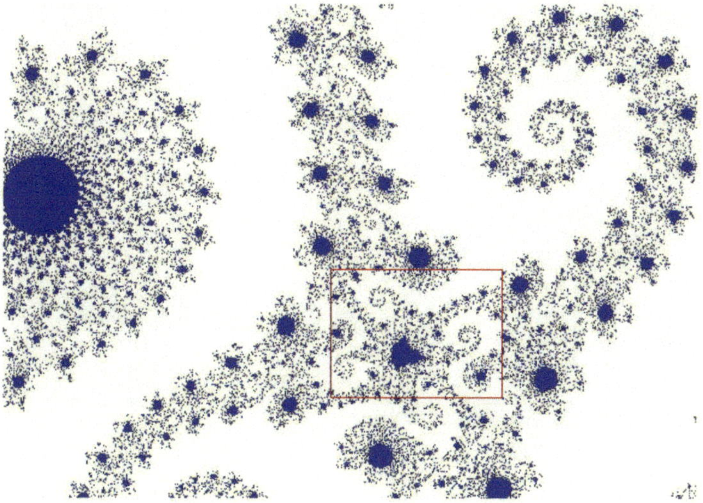

Figure 3.17. $0.08137 \leq x \leq 0.08141$, $0.649902 \leq y \leq 0.649932$; $r = 2$, $n = 500$. The red rectangle marks the area shown in Figure 3.18.

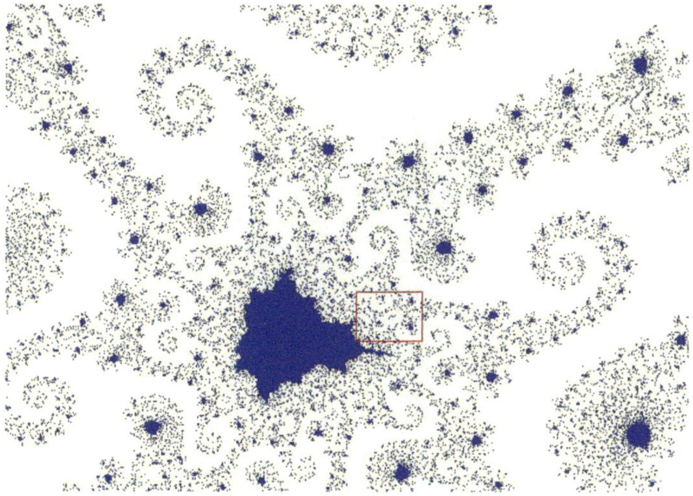

Figure 3.18. $0.081389 \leq x \leq 0.081398$, $0.6499092 \leq y \leq 0.649916$; $r = 2$, $n = 700$. The red rectangle marks the area shown in Figure 3.19.

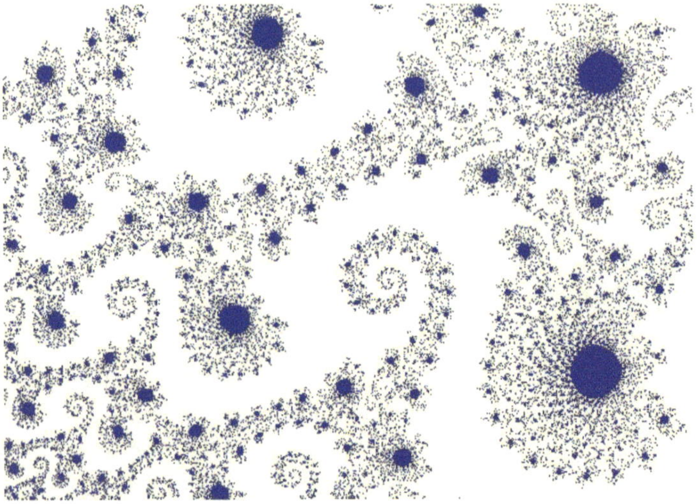

Figure 3.19. $0.0813936 \leq x \leq 0.0813944$, $0.6499115 \leq y \leq 0.6499121$; $r = 2$, $n = 700$.

The curiosity of the cows lined up at the frontier of the set $M(2, 100)$ (Figures 3.20, 3.21) seems to be concentrated on strange growths on the opposite slope of the trench between different buds as e.g. in Figures 3.22 and 3.24–3.26 (the last ones give an impression as if coming from somewhere in East Asia). The strange objects start to dissolve if the escape time bound $n = 100$ is increased to e.g. $n = 200$ (Figures 3.23 and 3.27), as has already happend with the bovine portraits in Figure 3.15.

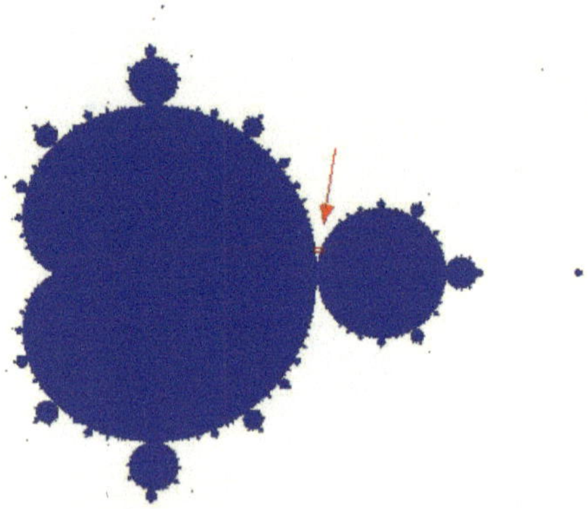

Figure 3.20. The red rectangle marks the area shown in Figure 3.21.

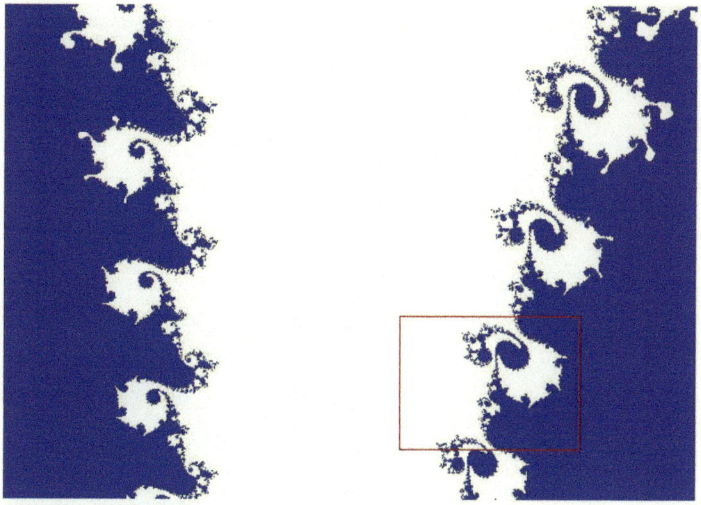

Figure 3.21. $0.74 \leq x \leq 0.77$, $0.0775 \leq y \leq 1.0000$; $r = 2$, $n = 100$. The red rectangle marks the area shown in Figure 3.22.

Figure 3.22. $0.757 \leq x \leq 0.764$, $0.08095 \leq y \leq 0.08620$; $r = 2$, $n = 100$.

Figure 3.23. Same limits as in Figure 3.22, but $r = 2$, $n = 200$.

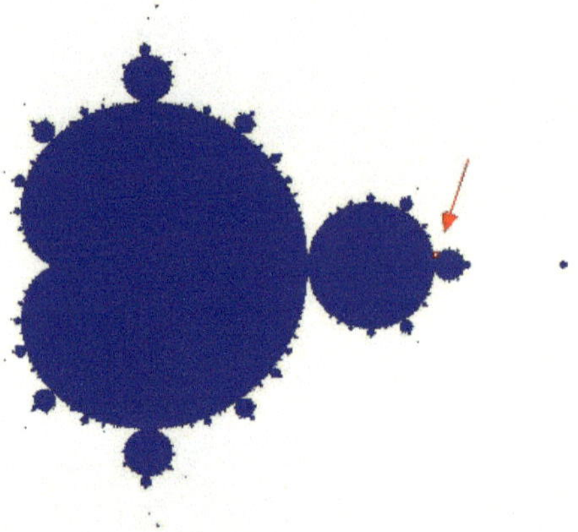

Figure 3.24. The red rectangle marks the area shown in Figure 3.25.

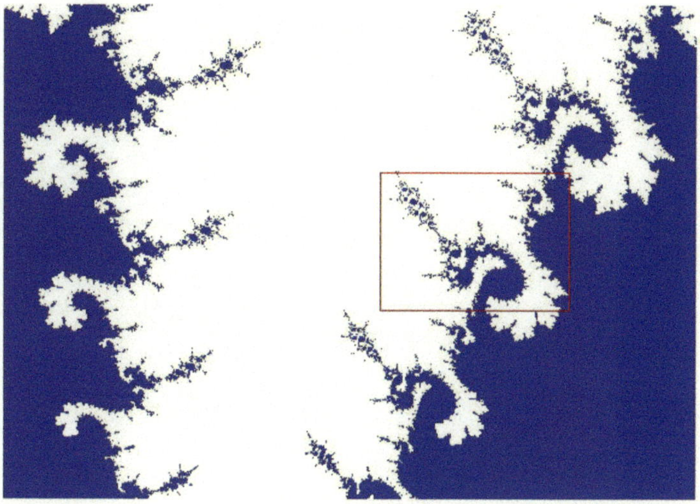

Figure 3.25. $1.2465 \leq x \leq 1.2650$, $0.030125 \leq y \leq 0.044000$; $r = 2$, $n = 100$. The red rectangle marks the area shown in Figure 3.26.

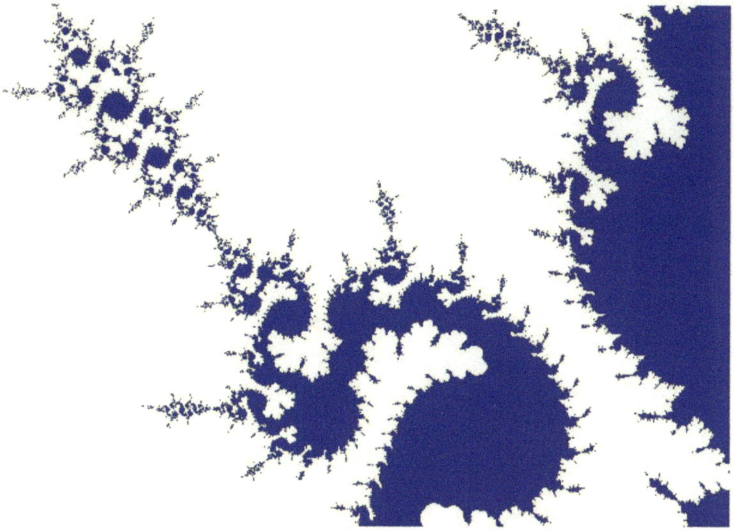

Figure 3.26. $1.2565 \leq x \leq 1.261$, $0.035625 \leq y \leq 0.039000$; $r = 10$, $n = 100$.

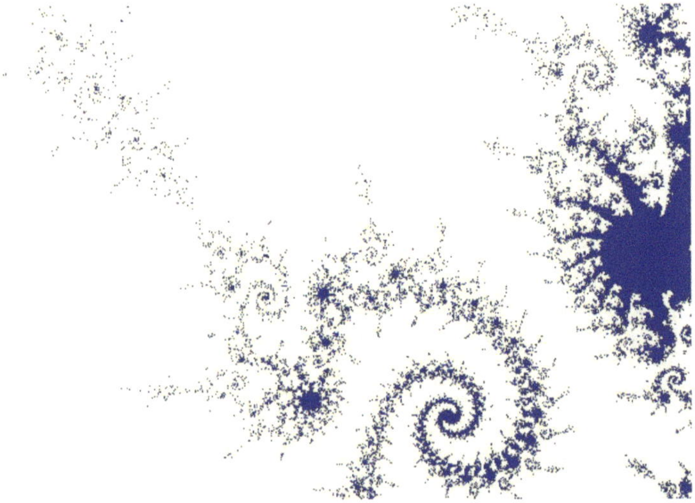

Figure 3.27. Same limits as in Figure 3.26, but now $r = 10$, $n = 200$.

We resist the temptation to dwell further on fantastic pictures provided lavishly by the boundary of M – an extensive sample is displayed in [Peitgen and Richter, 1986, maps 26–54]. But let us still explore at least two regions at the "outskirts" of M up north and in the east, where M ends at the point $c = 2$. The outer red region in Figure 3.3 informs us that the set M does not extend beyond $\mathbf{Im}\,(c) = 1.13$, and since the points of M actually depicted by pixels in this area seem to be sparse, we again resort to an escape time colour chart of the region indicated by the red rectangle in Figure 3.28 (Figure 3.29). Again – as in Figures 3.17 and 3.18 – we meet a curious feature of M: it seems to contain again and again smaller and smaller replicas of itself (commonly called *secondary* MANDELBROT *sets*). Up north an example is demonstrated in Figures 3.30 and 3.31, and in the far east (Figure 3.32) in Figure 3.33.

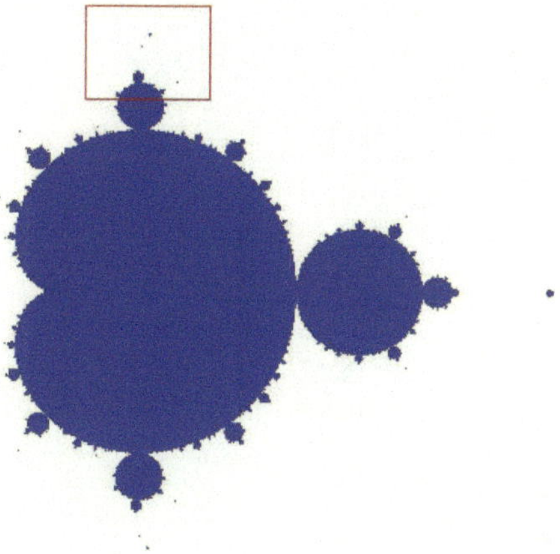

Figure 3.28. The red rectangle marks the area shown in Figure 3.29.

Figure 3.29. $-0.1 \le x \le 0.4$, $0.775 \le y \le 1.150$; $r = 2$, $n = 100$. The red rectangle marks the area shown in Figure 3.30.

Figure 3.30. $0.18 \leq x \leq 0.21$, $1.0875 \leq y \leq 1.1100$; $r = 2$, $n = 200$. The red rectangle marks the area shown in Figure 3.31.

Figure 3.31. $0.1973 \leq x \leq 0.2001$, $1.0993 \leq y \leq 1.1014$; $r = 2$, $n = 400$.

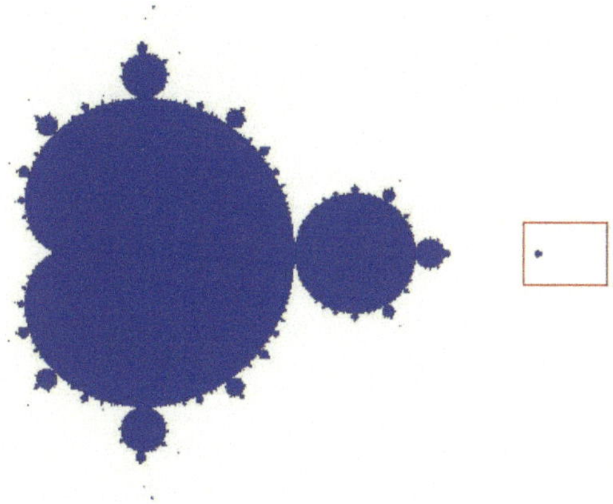

Figure 3.32. The red rectangle marks the area shown in Figure 3.33.

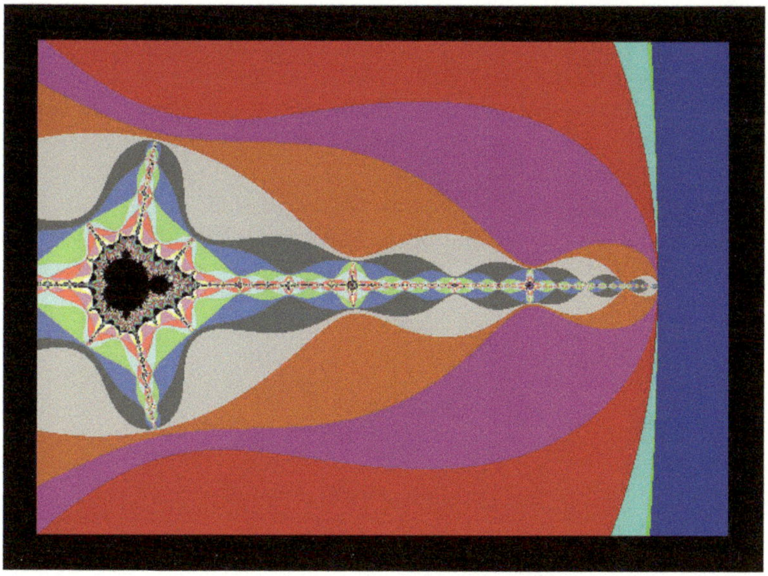

Figure 3.33. $1.70 \leq x \leq 2.05$, $-0.1325 \leq y \leq 1.3000$; $r = 2$, $n = 100$.

3.3.7 Regions with attractive periodic orbits

We already know that for $c \notin M$ there are no attractive periodic orbits of f (Theorem 3.1.22). So one gets curious about attractive periodic orbits in the case $c \in M$. There can be at most one by Theorem 3.1.20, and we have located the parameter values c for which there is an attractive fixed point or an attractive orbit of period 2 (Theorems 3.2.2 and 3.2.3). It becomes harder to do so for larger periods, but there is an odd way, avoiding any deep analysis, to obtain evidence about attractive periodic orbits for a given c with the help of a computer: start with an initial point z_0 "inside of J" (e.g. $z_0 = 0$) and let the computer hurry through the first, say, 10000 values of the orbit $\{z_k = f^{(k)}(z_0)\}_{k=0}^{\infty}$. By that time there is a good chance that this orbit will be so close to any attractive periodic orbit that the next values will practically coincide with it and signal a coincidence of the points $z_{10000+k} = z_{10000+k+p}$ after p further applications of f.

Let us try out this procedure first for real values of c and plot the (real) coordinates of the periodic points obtained in this way against the abscissas c on the real line. From Theorems 3.2.2 and 3.2.3 we already know that for $-0.25 < c < 0.75$ we should get a single value for an attractive fixed point; for $0.75 < c < 1.25$ we should jump to two values as members of an attractive periodic orbit of period 2, but what happens if we continue to send the value c further up to the last sensible value $c = 2$?

The picture produced by the computer (Figure 3.34) is somewhat surprising: as c increases, the number of periodic points is doubled again and again up to a certain point; there seemingly some chaos starts, but still a certain structure prevails. The phenomenon of this "bifurcation process" connected with period-doubling has been studied already by VERHULST (see [Verhulst, 1845], [Verhulst, 1847]) in the context of population dynamics (see also [May, 1976]). An attentive observer may recognize self-similarities in the produced graph which, due to the insight into the relevance of this phenomenon for dynamics revealed by [Feigenbaum, 1978], is also called a FEIGENBAUM *diagram*.

We still have no information about attractive periodic orbits with e.g. period 3. We can try the procedure having led up to Theorem 3.2.3: find roots of the equation $f^{(3)}(z) = z$.

$$
\begin{aligned}
f^{(3)}(z) &= \left[(z^2-c)^2-c\right]^2-c \\
&= z^8 - 4cz^6 + 2c(3c-1)z^4 - 4c^2(c-1)z^2 + c^4 - 2c^3 + c^2 - c \\
&= z.
\end{aligned}
$$

The equation $f^{(3)}(z) - z = 0$ is certainly satisfied by every fixed point, so the left side must be divisible by $(z^2 - z - c)$. Indeed we have

$$
\begin{aligned}
&z^8 - 4cz^6 + 2c\,(3c-1)z^4 - 4c^2(c-1)z^2 - z + c\,(c^3 - 2c^2 + c - 1) \\
&\quad = (z^2 - z - c)\cdot\left[z^6 + z^5 + (1-3c)z^4 + (1-2c)z^3 + (1-3c+3c^2)z^2\right. \\
&\qquad \left. + (1-2c+c^2)z + 1 - c + 2c^2 - c^3\right].
\end{aligned}
\tag{3.7}
$$

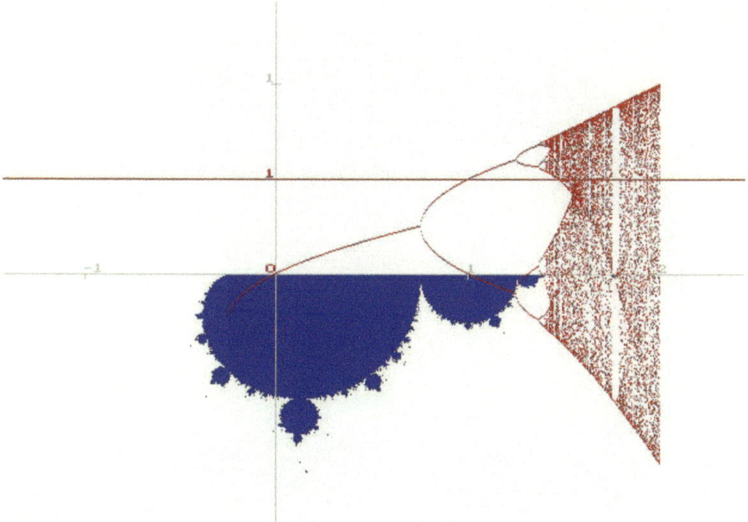

Figure 3.34. For values of $c \in [-\frac{1}{4}, 2]$ on the real axis there are real non-repellent periodic orbits $\{x_k = f^{(k)}(x_0)\}_{k=0}^{p-1}$; the graph in brown displays the corresponding values $-x_k$ $(0 \leq k < p)$ (measured with the brown line segment $\overline{01}$ as unit) as ordinates.

If w_1 is a root of the polynomial (3.7), then so are $w_2 = f(w_1)$ and $w_3 = f(w_2)$, and $\{w_1, w_2, w_3\}$ must be an orbit of period 3. But there must be six roots, and since in general the polynomial (3.7) will not be the square of a cubic polynomial, there must be another orbit $\{w_4,\ w_5 = f(w_4),\ w_6 = f(w_5)\}$ of period 3. The solution of this, at a first glance maybe puzzling, situation is that in general there are two different periodic orbits with period 3 belonging to the JULIA set corresponding to the parameter c, at most one of them being attractive. Since by Theorem 3.1.4,

$$f^{(3)'}(w_1) = 8w_1w_2w_3, \qquad f^{(3)'}(w_4) = 8w_4w_5w_6,$$

and

$$\prod_{k=1}^{6} w_k = 1 - c + 2c^2 - c^3,$$

this is certainly the case if

$$1 - c + 2c^2 - c^3 = 0. \tag{3.8}$$

In this case the attractive orbit is even super-attractive since it must contain the point 0; consequently it has the form $\{0, -c, c(c-1)\}$. The three solutions of (3.8) are

$$c_{1,2} \approx 0.12256117 \pm 0.74486177\, i, \qquad c_3 \approx 1.75487767.$$

The JULIA set corresponding to c_3 is not spectacular (Figure 3.35). The conjugate complex parameters c_1 and c_2 lie in the big buds at the top and the bottom of the MANDELBROT set. The corresponding JULIA sets (Figure 3.36) for c_1 also shows an orbit

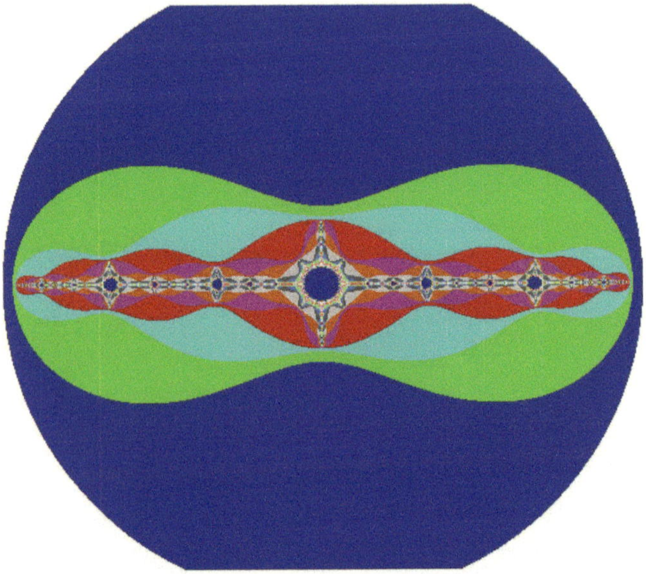

Figure 3.35. The escape time colour chart for the JULIA set corresponding to $c = 1.75487767$ for $r = 2$, $n = 200$. The super-attractive orbit of period 3 consists of the points 0, $-c$ and $c(c-1) \approx 1.32471797$.

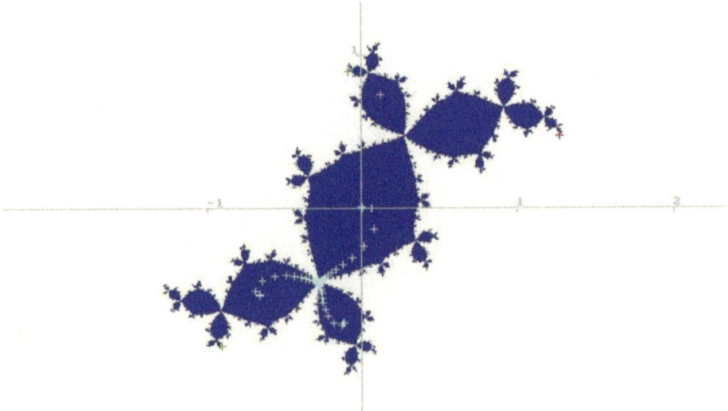

Figure 3.36. The escape time approximation of DOUADY's rabbit, i.e. for the filled JULIA set corresponding to $c \approx 0.12256117 + 0.74486177\,i$ (marked by a grey cross) for $r = 2$, $n = 200$. The super-attractive orbit of period 3 consists of the points 0, $-c$, and $c(c-1) \approx -0.66236 - 0.56228\,i$. The orbit marked by light cyan crosses starts close to the repellent fixed point $z_1 \approx -0.2763 - 0.4797\,i$. The second repellent fixed point (marked by a red cross) is $z_2 \approx 1.2763 + 0.4797\,i$, the periodic orbit with period 2 consists of the points $z_3 \approx -0.0838 + 0.8948\,i$ and $z_4 \approx -0.9162 - 0.8948\,i$ (marked by green crosses).

converging to the super-attractive orbit with period 3) are called DOUADY*'s rabbit.* A continuous change of the parameter c should not change much of the situation, so there ought to be an open set, containing c_1, of parameter values c which admit an attractive periodic orbit with period 3, and the bud of M containing c_1 seems to be a good candidate for it.

It would be tiresome to pursue in general the same procedure in the search for attractive periods of period $k \in \mathbb{N}$. There is a heuristic reasoning shedding some light on the location of the corresponding values of c: Suppose $\frac{l}{k}$ is a reduced fraction, i.e. $(l, k) \in \mathbb{Z} \times \mathbb{N}$ and l and k are relatively prime; if $\varphi_{l,k} = \frac{2\pi l}{k}$ and $c_{l,k} = \frac{1}{4}(e^{2i\varphi_{l,k}} - 2e^{i\varphi_{l,k}})$, then Theorem 3.2.2 informs us that the point $z_1 = \frac{e^{2\pi li/k}}{2}$ is a neutral fixed point with $f'(z_1) = e^{2\pi li/k}$. For $\frac{l}{k} = -\frac{1}{3}$ this limiting case is illustrated in Figure 3.37. The action of f in the neighbourhood of z_1 consists approximately in a rotation about the angle $\frac{2\pi l}{k}$ which tries hard to produce periodic orbits of period k. Traversing the cycloid C_0 at $c_{l,k}$ to the outside of it makes z_1 repellent and hands attractiveness indeed over to one of the orbits of period k, a situation already illustrated in Figure 3.36. In fact it turns out that the points $c_{l,k}$ are the points at which buds are attached to C_0, for $l = \pm 1$ decreasing in size as $k \to \infty$, which harbor parameter values associated with JULIA sets admitting attractive periodic orbits with period k. An example for $k = 20$ is depicted in Figure 3.38. There are more laws governing the army of buds connected directly or via other buds to the interior of the cycloid C_0, but we leave the mathematics behind this structure to [Blanchard, 1984] and the sources cited there, as well as the assertions justifying the subsumption of the MANDELBROT and JULIA sets under the family of fractals and identifying their HAUSDORFF dimension.

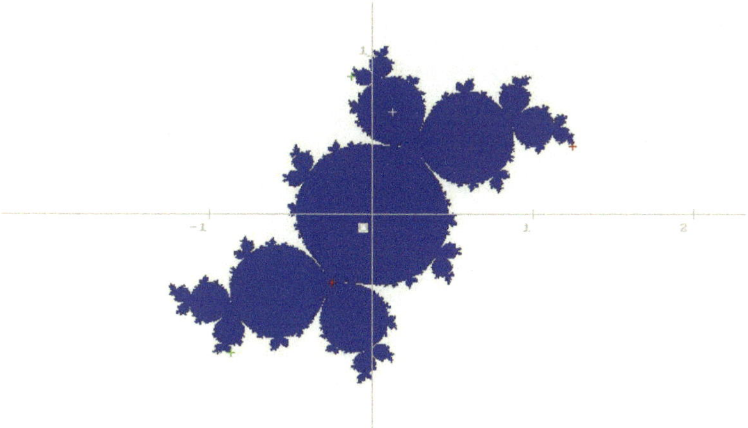

Figure 3.37. The escape time approximation of the filled JULIA set corresponding to $c_{-1,3} = \frac{1}{4}(e^{-4\pi i/3} - 2e^{-2\pi i/3}) \approx 0.125 + 0.6495\,i$ on the cycloid C_0, for $r = 1.5$, $n = 1000$. The point $z_1 = e^{-2i\pi/3} \approx -0.2500 - 0.4330\,i$ is a neutral fixed point, the point $z_2 \approx 1.2500 + 0.4330\,i$ a repellent one (both marked by red crosses). The periodic orbit with period 2 consists of the points $z_3 \approx -0.1283 + 0.8736\,i$ and $z_4 \approx -0.8717 - 0.8736\,i$ marked by green crosses.

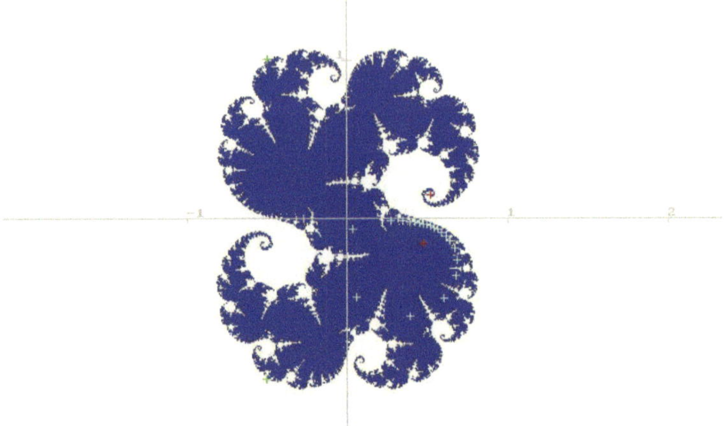

Figure 3.38. The escape time approximation of the filled JULIA set corresponding to $c = -0.2733 + 0.0074\,i$ for $r = 2$, $n = 200$. The two fixed points $z_1 \approx -0.4761 - 0.1545\,i$, $z_2 = 0.5239 + 0.1545\,i$, marked by red crosses, are repellent, if also not very strongly: $|f'(z_1)| \approx 1.0010$, $|f'(z_2)| \approx 1.0925$. The two points $z_3 \approx -0.4963 + 1.0116\,i$ and $z_4 \approx -0.5037 - 1.0116\,i$, marked by green crosses, constitute the periodic orbit with period 2 and $|{f^{(2)}}'(z_3)| = |{f^{(2)}}'(z_4)| \approx 5.0933$. There is an attractive periodic orbit with period 20, marked by light cyan crosses, one point of which is $\approx 0.27066 - 0.01203\,i$.

In the illustrations of JULIA sets to follow, as already done in Figures 3.36–3.38, the fixed points will be marked by red crosses and the points of the orbit of period 2 by green crosses, while the location of c will be marked by a grey cross.

3.4 Generation of JULIA sets

3.4.1 The sets $J_c(r, n)$

We have already used the method of escape time and corresponding colour charts in order to construct with the help of a computer an – approximate – picture of the JULIA set $J = J_c$ for the function $f = f_c$ in (3.4). As done with the MANDELBROT set in the beginning of Section 3.3.4, in practice we are mostly bound to be content with a larger set $J(r, n) = J_c(r, n) = \{z : |f_c^{(k)}(z)| \le r\ \forall\ k \le n\}$. This set might contain some point (or even an open set) where there is none in J, depicted e.g. in blue, but it is a mathematically well defined, bona fide set closely related to the JULIA set J (since $J(r, n) \supset J(r, n+1)$ and for the filled JULIA set $\tilde{J}$ we have $\tilde{J} = \bigcap_{n=1}^{\infty} J(r, n)$) and sometimes aesthetically pleasing. So we may legitimately be interested in such sets, examples of which are given in Figures 3.39–3.43. Again, a beautiful collection of JULIA sets is presented in [Peitgen and Richter, 1986].

There are at least two other ways to approximately depict JULIA sets, suggested by what we have found out in Section 3.1 and Section 3.2.

Figure 3.39. The set $J(2, 100)$ for $c = 0.7454 + 0.1130\,i$. Fixed points are $z_1 \approx -0.4993 - 0.0565\,i$ (almost neutral) and $z_2 \approx 1.4993 + 0.0565\,i$. The periodic orbit with period 2 consists of the points $z_3 \approx -0.2670 + 0.2426\,i$ and $z_4 \approx -0.7330 - 0.2426\,i$.

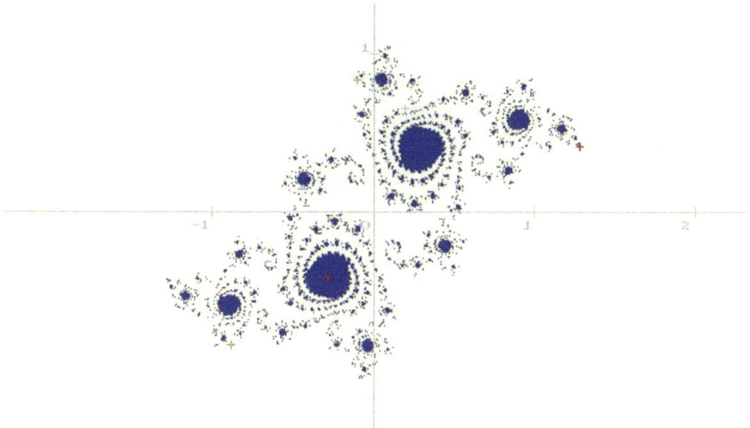

Figure 3.40. The set $J(2, 100)$ for $c = 0.1940 + 0.6557\,i$. Fixed points are $z_1 \approx -0.2861 - 0.4171\,i$ (almost neutral) and $z_2 \approx 1.2851 + 0.4171\,i$. The periodic orbit with period 2 consists of the points $z_3 \approx -0.1103 + 0.8413\,i$ and $z_4 \approx -0.8897 - 0.8413\,i$.

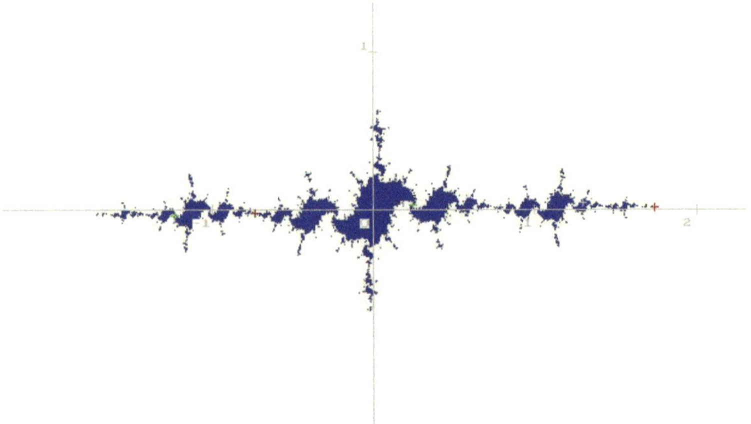

Figure 3.41. The set $J(2, 100)$ for $c = 1.3000+0.0500\,i$. Fixed points are $z_1 \approx -0.7452 - 0.0201\,i$ and $z_2 \approx 1.7452 + 0.0201\,i$. The periodic orbit with period 2 consists of the points $z_3 \approx -1.2424 + 0.0337\,i$ and $z_4 \approx 0.2424 + 0.0337\,i$.

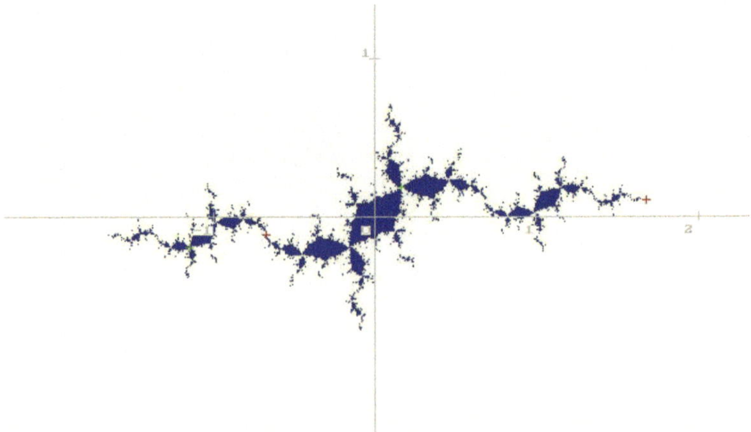

Figure 3.42. The set $J(2, 100)$ for $c = 1.1500+0.2500\,i$. Fixed points are $z_1 \approx -0.6879 - 0.1052\,i$ and $z_2 \approx 1.6879 + 0.1052\,i$. The periodic orbit with period 2 consists of the points $z_3 \approx -1.1602 + 0.1893\,i$ and $z_4 \approx 01.1602 + 0.1893\,i$.

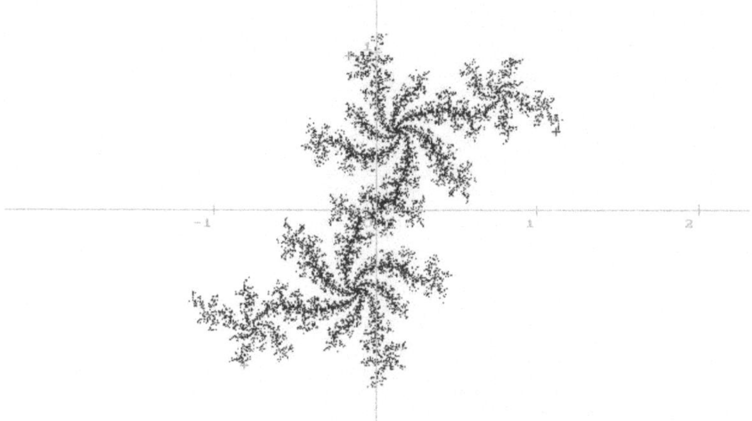

Figure 3.43. The set $J(2, 100)$ for $c = -0.1103 + 0.6300\,i$. Fixed points are $z_1 \approx -0.1265 - 0.5028\,i$ and $z_2 \approx 1.1265 + 0.5028\,i$. The periodic orbit with period 2 consists of the points $z_3 \approx -0.8209 - 0.9815\,i$ and $z_4 \approx -0.1791 + 0.9815\,i$.

3.4.2 Pre-images of repellent periodic orbits

Theorem 3.2.2 furnishes at least one repellent fixed point z_2 for f which certainly belongs to $J(f)$. Since $J(f)$ is forward and backward invariant (Theorem 3.1.13), it also contains all inverse image sets $f^{(-k)}(\{z_2\})$ containing 2^k points which again can be visualized on the screen of a computer. In fact, every member of a repellent periodic orbit would do as well. The good news is that the set $\bigcup_{k=0}^{\infty} f^{(-k)}(\{z_2\})$ fills up $J(f)$ densely from the inside by Theorem 3.1.26, but the bad news is that this may happen in a rather irregular way, preferring certain parts of $J(f)$ (where then a whole bunch of points of $J(f)$ are represented by a single pixel), and for a long time rather neglecting other parts of $J(f)$.

For parameter values $c \in M$ this procedure has the advantage to show the proper JULIA set, not only as the boundary of the set inside of it – at least as long as the last mentioned drawback remains bearable (Figure 3.44).

Still, also in this case, an escape time picture may appear more satisfactory than the backward orbit of a fixed point which can be deceptive if parts of J are visited by pre-images too sparsely (Figures 3.45–3.48).

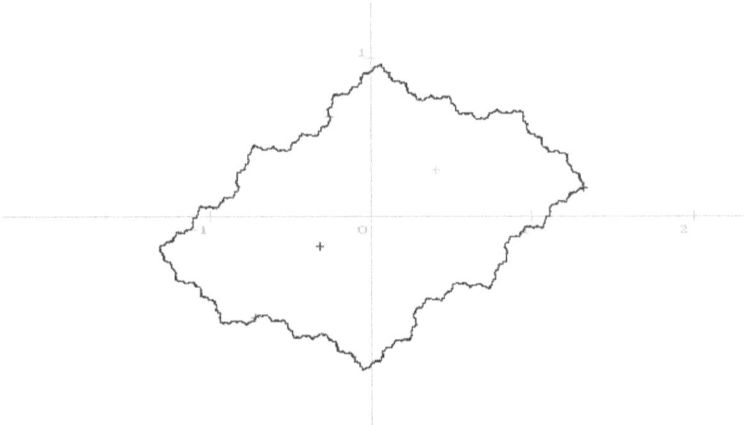

Figure 3.44. The set $f^{(-14)}(\{z_2\})$ for $c = 0.4 + 0.3\,i$. Fixed points are $z_1 \approx -0.3264 - 0.1815\,i$ (attractive) and $z_2 \approx 1.3264 + 0.1815\,i$ (repellent). The periodic orbit with period 2 consists of the points $z_3 \approx -0.2644 + 0.6368\,i$ and $z_4 \approx -0.7356 - 0.6368\,i$.

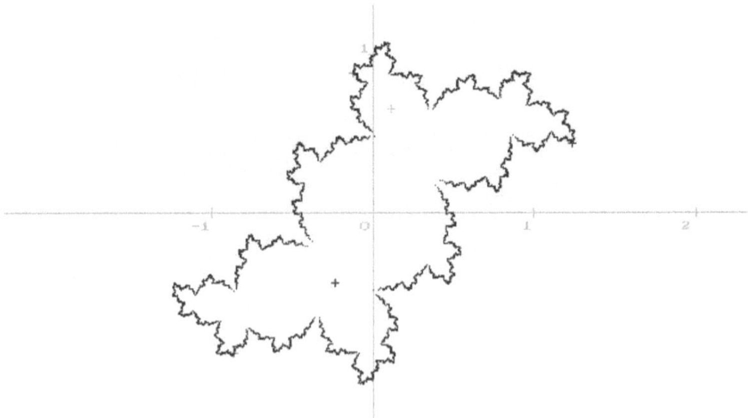

Figure 3.45. The set $f^{(-17)}(\{z_2\})$ for $c = 0.1100 + 0.6557\,i$. Fixed points are $z_1 \approx -0.2443 - 0.4405\,i$ (almost neutral) and $z_2 \approx 1.2443 + 0.4405\,i$ (repellent). The periodic orbit with period 2 consists of the points $z_3 \approx -0.1283 + 0.8821\,i$ and $z_4 \approx -0.8717 - 0.8821\,i$.

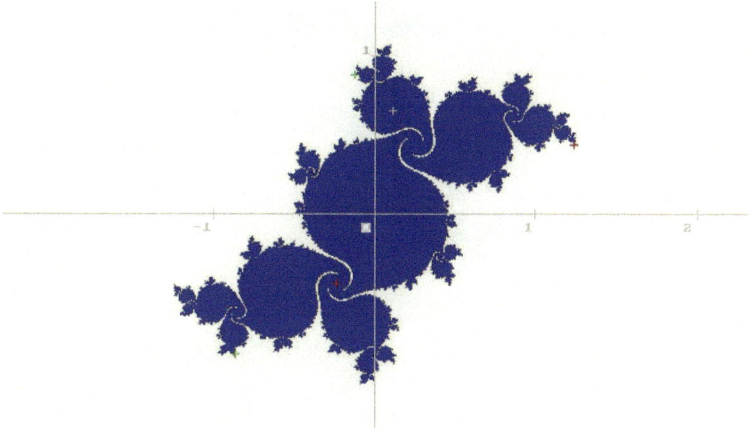

Figure 3.46. The set $J(2, 200)$ for $c = 0.11000 + 0.6557\,i$ as in Figure 3.45.

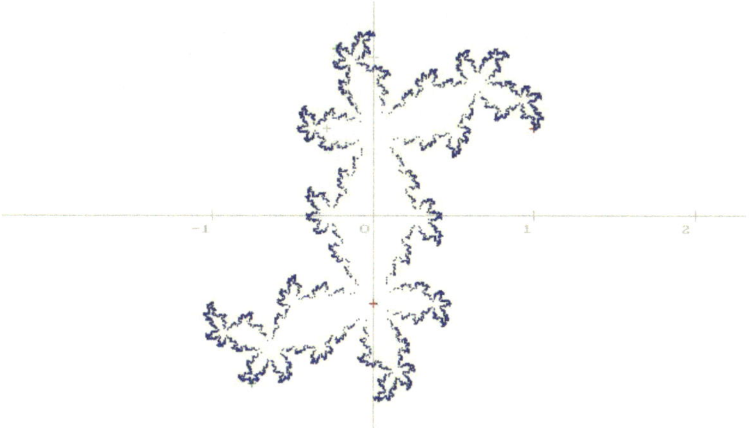

Figure 3.47. The set $f^{(-18)}(\{z_1\})$ for $c = -0.3000 + 0.5500\,i$. Fixed points are $z_1 \approx -0.0011 - 0.5488\,i$ and $z_2 \approx 1.0011 + 0.5488\,i$ (both repellent). The periodic orbit with period 2 consists of the points $z_3 \approx -0.7601 - 1.072\,i$ and $z_4 \approx -0.2399 + 1.0572\,i$.

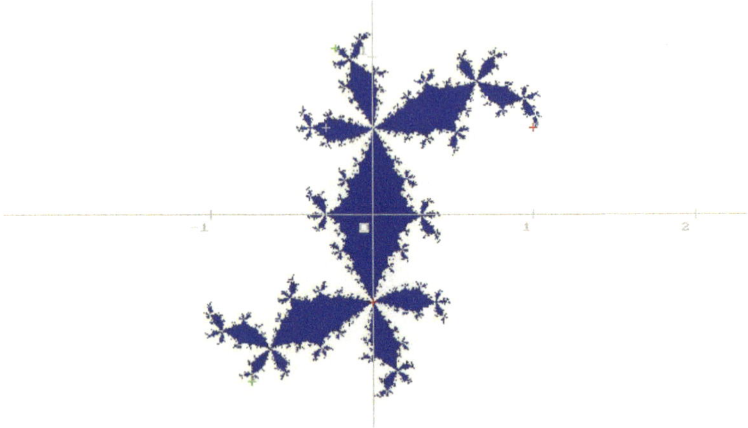

Figure 3.48. The set $J(2, 200)$ for $c = -0.3000 + 0.5500\,i$ as in Figure 3.47.

For parameter values $c \notin M$ the graph of the backward orbit of a fixed point may give a more informative picture than the escape time (Figures 3.49, 3.50) or at least help to interpret better an escape time colour chart (Figures 3.51, 3.5, 3.6); it may even appear more satisfactory than an escape time picture (Figures 3.52, 2.5). So this procedure – theoretically perfect – with all of its shortcomings in practice appears certainly to be useful.

Figure 3.49. The set $f^{(-20)}(\{z_2\})$ for $c = 0.8\,i$. This parameter value does not belong to M as demonstrated in Figure 3.50. Fixed points are $z_1 \approx -0.2376 - 0.5423\,i$ and $z_2 \approx 1.2376 + 0.5423\,i$. The (repellent) periodic orbit with period 2 consists of the points $z_3 \approx -0.9163 - 0.9609\,i$ and $z_4 \approx -0.0837 + 0.9609\,i$.

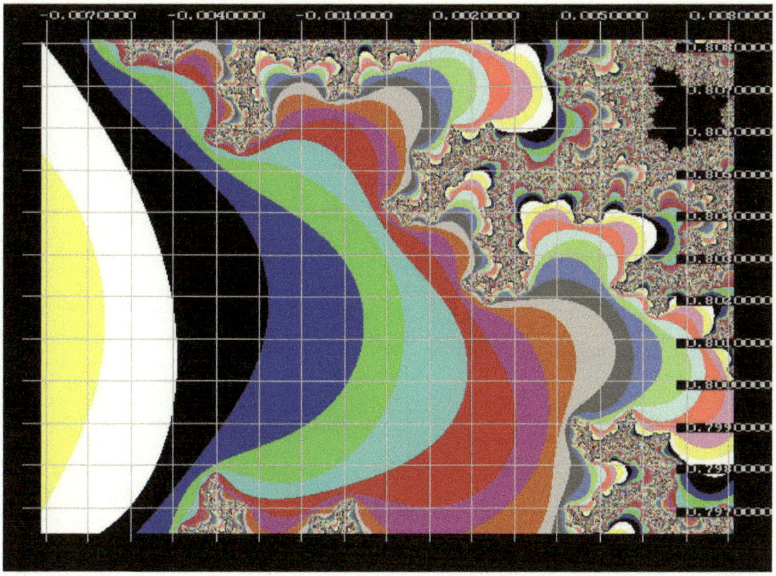

Figure 3.50. An escape time colour chart of the subset of the MANDELBROT set for $-0.009 \leq x \leq 0.009$, $0.796 \leq y \leq 0.809$, $r = 2$, $n = 200$. The point $c = 0.8\,i$ (Figure 3.49) belongs to the green escape-area in the center of the rectangle; in fact one has $|f_c^{(18)}(0)| > 2$.

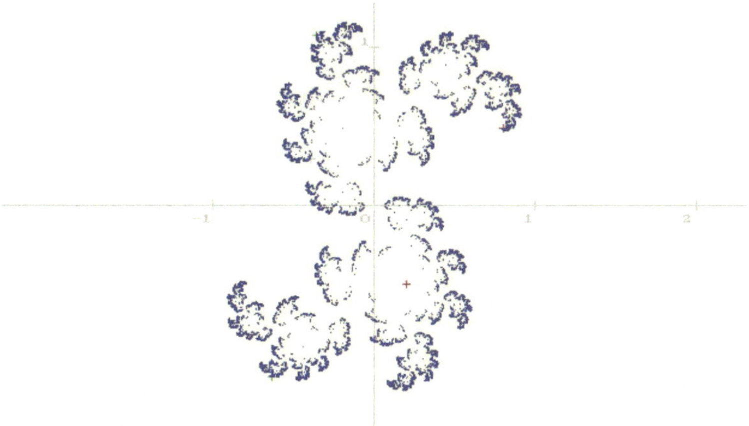

Figure 3.51. The set $f^{(-20)}(\{z_1\})$ for $c = -0.4000 + 0.3000\,i$ as in Figures 3.5 and 3.6.

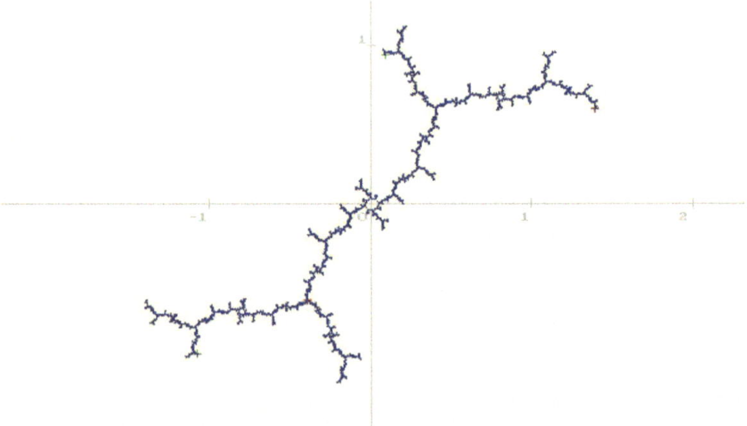

Figure 3.52. The set $f^{(-17)}(\{z_1\})$ for $c = 0.1981 + 1.1002\,i$ in the center of the secondary MANDELBROT set in Figure 3.31. Fixed points are $z_1 \approx -0.4044 - 0.6082\,i$ and $z_2 \approx 1.4044 + 0.6082\,i$ (both repellent). The periodic orbit with period 2 consists of the points $z_3 \approx -1.0827 - 0.9441\,i$ and $z_4 \approx 0.0827 + 0.9441\,i$.

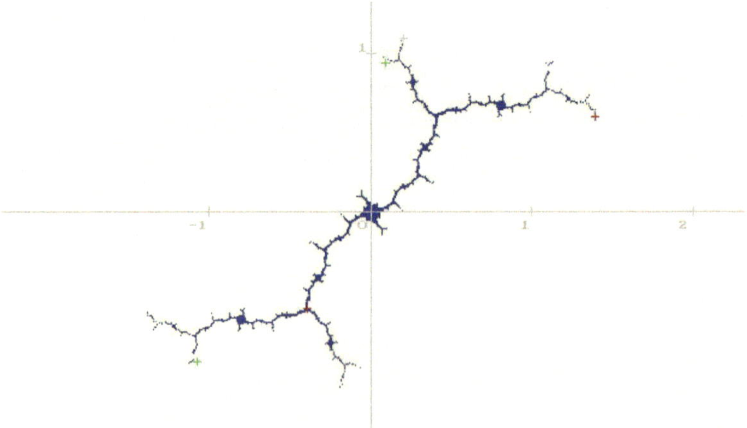

Figure 3.53. The set $J(2, 13)$ for $c = 0.1981 + 1.1002\,i$ as in Figure 3.52.

3.4.3 The JULIA set as attractor

Lemma 3.1.10 also asserts that the exterior E of the closed disc $B(0, r_0)$ is mapped by f into the exterior of $B(0, 2r_0)$. As a consequence we have $f(B(0, r_0)) \supset B(0, r_0)$ and $f^{(-1)}(B(0, r_0)) \subset B(0, r_0)$. In other words, on $B(0, r_0)$ the inverse map $f^{(-1)}$ behaves like a contraction and $\bigcap_{k=1}^{\infty} f^{(-k)}(B(0, r_0)) = \mathbb{C} \setminus \bigcup_{k=1}^{\infty} f^{(-k)}(E) = \mathbb{C} \setminus A(f, \infty)$ (in fact, every disc containing J in its interior would do as well). Therefore, constructing the intersection produces a set with boundary $J(f)$. This construction is still supported e.g. if, in case there exists an attractive fixed point z_1, a sufficiently small disc $B(z_1, r_1)$ with center z_1, lying inside of the connected set J, is deleted from $B(0, r_0)$ (Figure 3.54): since f works as a contraction in $A(f, z_1)$ with $J = \partial A(f, z_1)$, the inverse map $f^{(-1)}$ contracts the set $B(0, r_0) \setminus B(z_1, r_1)$ eventually to J.

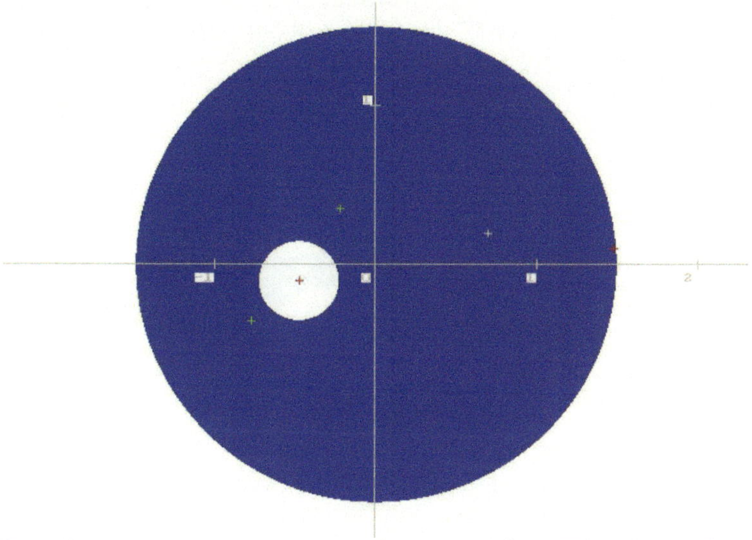

Figure 3.54. The initial set F, the complement of $B(z_1, 0.25)$ in $B(0, 1.5)$, for the contracting inverse map ${f_c}^{(-1)}$ for $c = 0.7 + 0.2\,i$ (marked by a grey cross) and the attractive fixed point $z_1 \approx -0.4800 - 0.1020\,i$.

This construction at the same time gives a supplementary answer to the question why a JULIA set should be considered as a fractal. It is the attractor of a sequence of sets formed by applying the iterates of the map $f^{(-1)}$ to a set $F \supset J$. This looks very much like an iterative function system since $f^{(-1)}(F)$ is produced by two functions defined by $g_1(w) = \sqrt{w + c}$ and $g_2(w) = -\sqrt{w + c}$ (after having chosen a branch of the complex square root) applied to F, and f as well as $f^{(-1)} = g_1 \cup g_2$ furnish conformal mappings. However, the two mappings g_j are not contractions in the sense of Definition 2.2.1 since e.g. $\sqrt{0.25} - \sqrt{0.16} = 0.5 - 0.4 = 0.1 > 0.25 - 0.16$. As a practical consequence, in the neighbourhood of $-c$ the images of w under g_1 and g_2 expand and in the screen graphics there appear lines of pixels not carrying images of points of the set F (Figures 3.55 and 3.61), which lines are proliferated through the further applications of $f^{(-1)}$ (Figures 3.56, 3.57, 3.62).

Figure 3.55. The set $f^{(-1)}(F)$ for c as in Figure 3.54. The functions g_1 and g_2 constituting the inverse map $f^{(-1)}$ are chosen so as to provide the square roots $\sqrt{w+c}$ with non-negative resp. non-positive real part. The white lines signal the lack of image-pixels as $f^{(-1)}$ is applied to the neighbourhood of $w = -c$.

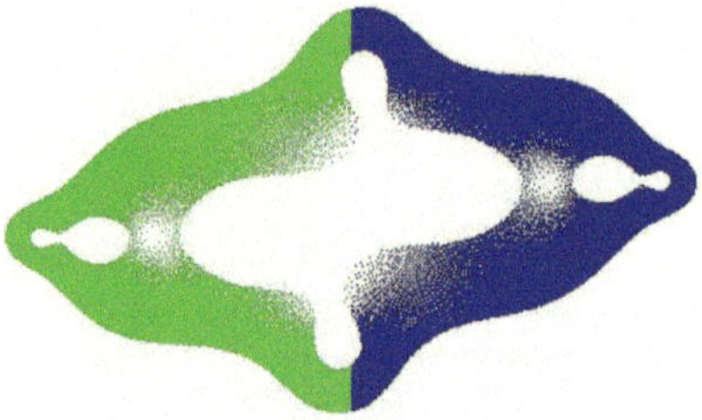

Figure 3.56. The set $f^{(-3)}(F)$ for c as in Figure 3.54.

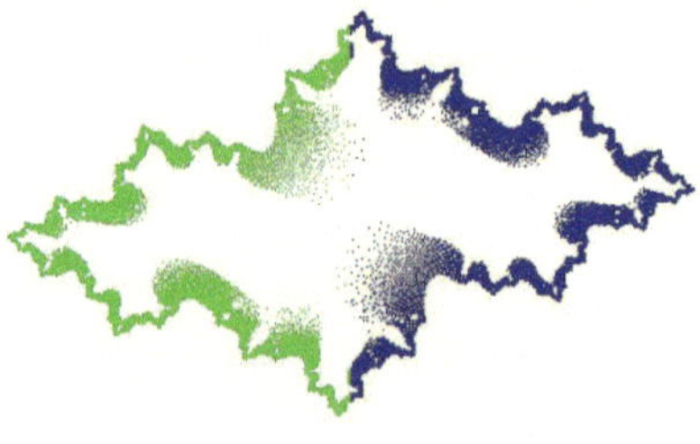

Figure 3.57. The set $f^{(-8)}(F)$ for c as in Figure 3.54.

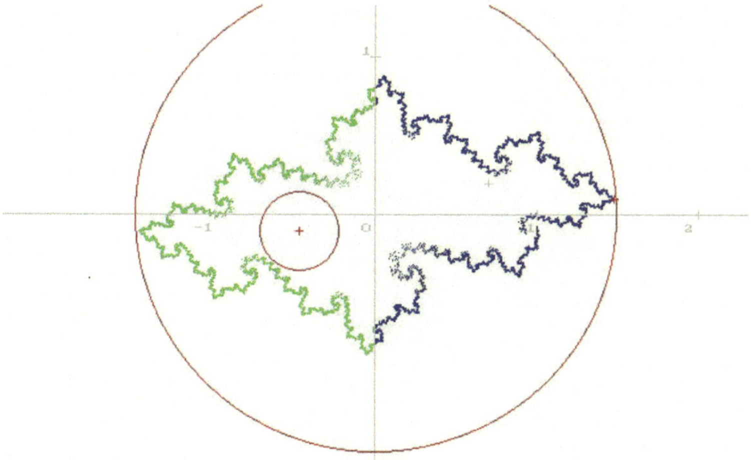

Figure 3.58. The set $f^{(-47)}(F)$ for c as in Figure 3.54. The repellent fixed point is $z_2 \approx 1.4800 + 0.1020\,i$, the periodic orbit of period 2 consists of the points $z_3 \approx -0.7794 - 0.3579\,i$, $z_4 \approx -0.2206 + 0.3579\,i$. The circles delimit the original set F.

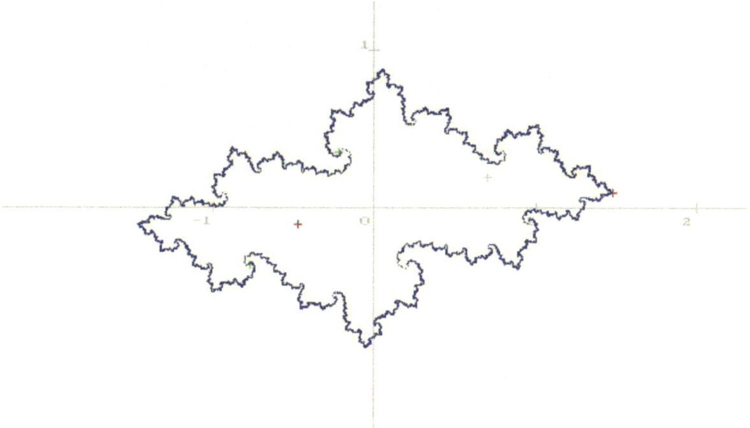

Figure 3.59. The set $f^{(-21)}(\{z_2\})$ for c as in Figures 3.54–3.58 and z_2 as in Figure 3.58.

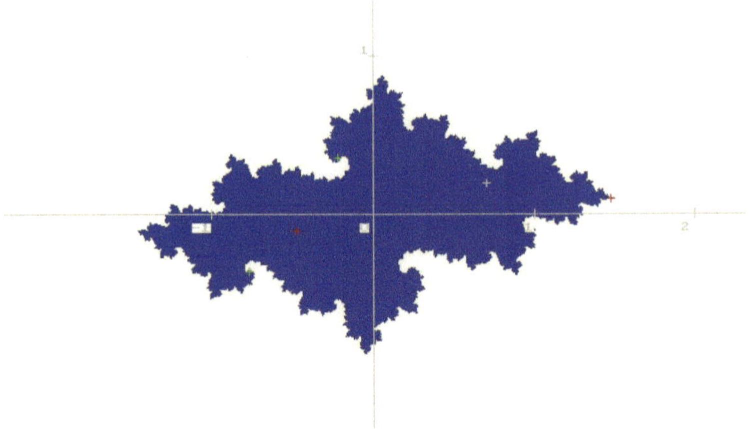

Figure 3.60. The set $J(2, 20)$ for c as in Figures 3.54–3.59.

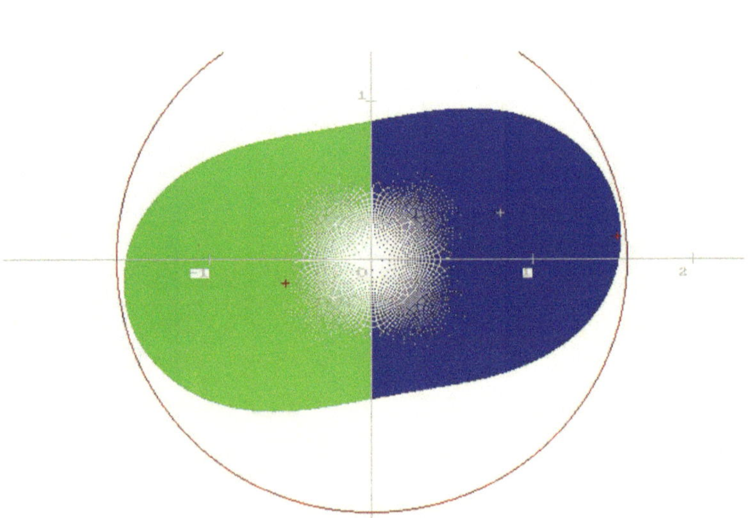

Figure 3.61. The set $f^{(-1)}(F)$ for $c = 0.8 + 0.3\,i \notin M$ and $F = B(0, 1.6)$ with red circumference. The repellent fixed points are $z_1 \approx -0.5349 - 0.1449\,i$, $z_2 \approx 1.5349 + 0.1449\,i$. The periodic orbit of period 2 consists of the points $z_3 \approx -0.9208 - 0.3565\,i$, $z_4 \approx -0.0792 + 0.3565\,i$.

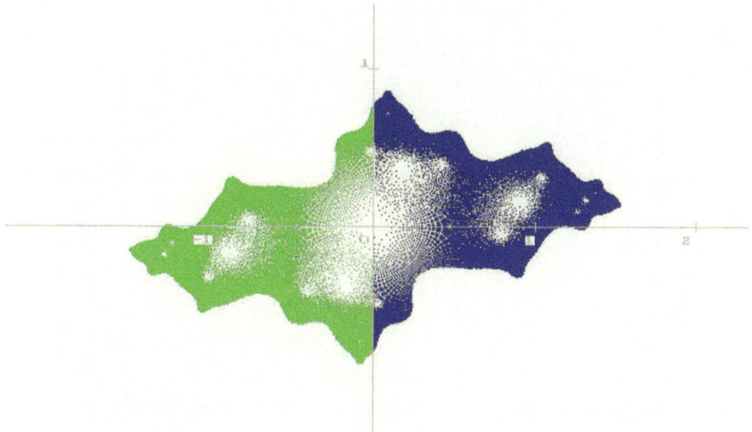

Figure 3.62. The set $f^{(-5)}(F)$ for c and F as in Figure 3.61.

A comparison of the various methods of approximately computer-picturing a JULIA set may be carried out on the basis of Figures 3.58–3.60 in case of $c \in M$ and of Figures 3.63 and 3.64 in case of $c \notin M$.

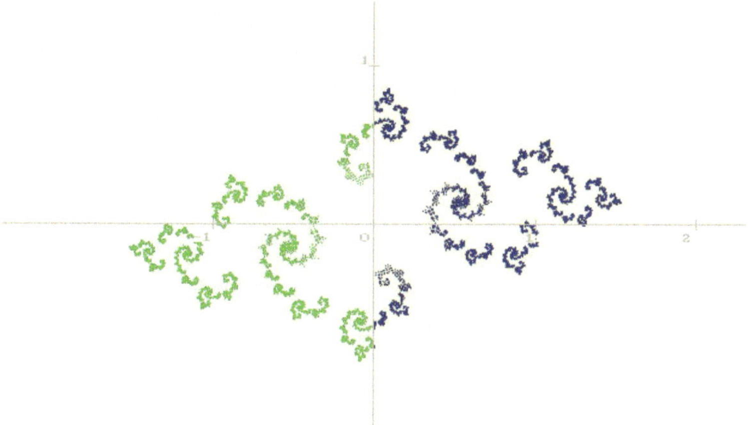

Figure 3.63. The set $f^{(-30)}(F)$ for c and F as in Figure 3.61.

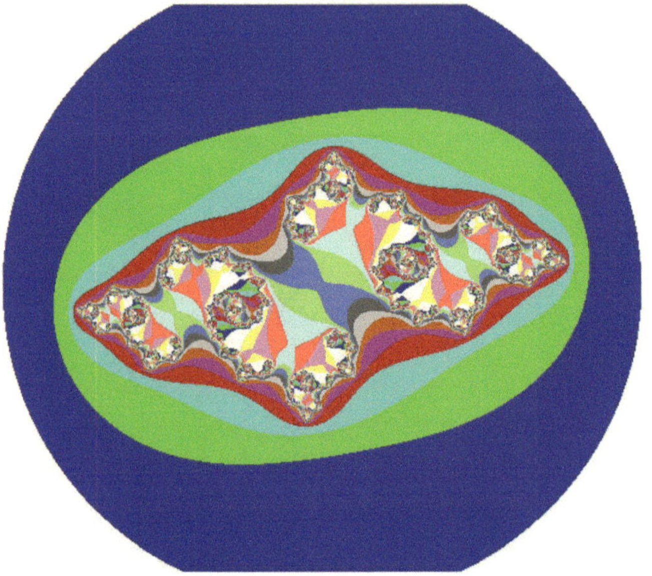

Figure 3.64. An escape time colour chart for the set $J(2, 200)$ and c as in Figures 3.61–3.63.

Good luck now for playing with fractals, and for learning more about them in the references cited below!

Bibliography

The pages on which the author is cited are indicated in italics. Much more literature on fractals is listed in the references below, in particular in [Falconer, 1990], [Mandelbrot, 1982], [Peitgen, Jürgens, Saupe, 1992].

Literature on fractals

[Addison, 1997] P. S. Addison. *Fractals and Chaos: An Illustrated Course.* Institute of Physics Publishing, Bristol, 1997.

[Barnsley, 1988] M. Barnsley. *Fractals Everywhere.* Academic Press, Inc., Toronto, 1988. (*v, 63, 85, 97, 98, 105*)

[Becker and Dörfler, 1989] K.-H. Becker and M. Dörfler. *Dynamical Systems and Fractals. Computer Graphics Experiments in Pascal.* Cambridge University Press, Cambridge, 1989. German Edition: *Computergraphische Experimente in Pascal.* Vieweg, Braunschweig 1986.

[Blanchard, 1984] P. K. Blanchard. *Complex analytic dynamics on the Riemann sphere.* Bull. Am. Math. Soc., 11:85–141, 1984. (*111, 117, 121, 131, 149*)

[Cantor, 1883] G. Cantor. *Über unendliche, lineare Punktmannigfaltigkeiten, V.* Math. Ann., 21:545–591, 1883. (*2*)

[Collet and Eckmann, 1980] P. Collet and J.-P. Eckmann. *Iterated Maps of the Interval as Dynamical Systems.* Birkhäuser, Basel, 1980.

[Davis and Knuth, 1970] C. Davis and D. E. Knuth. *Number representations and dragon curves.* Journal of Recreational Mathematics, 3:66–81, 133–149, 1970. (*27*)

[Douady and Hubbard, 1982] A. Douady and J. H. Hubbard. *Itération des polynômes quadratiques complexes.* C. R. Acad. Sci., Paris, Sér. I, 294:123–126, 1982. (*126*)

[Douady, 1983] A. Douady. *Systèmes dynamiques holomorphes.* Séminaire Bourbaki, 35e année, 1980/81, Exp. 599, Astérisque, 105–106:39–63, 1983. (*117, 148, 149*)

[Falconer, 1985] K. J. Falconer. *The Geometry of Fractal Sets.* Cambridge University Press, Cambridge, 1985.

[Falconer, 1990] K. J. Falconer. *Fractal Geometry. Mathematical Foundations and Applications.* Wiley, New York, 1990. German Edition: *Fraktale Geometrie.* Spektrum Akademischer Verlag, Heidelberg, 1993. (*v, 51, 60, 111*)

[Feder, 1988] P. Feder. *Fractals.* Plenum Press, New York, 1988.

[Frame and Mandelbrot, 2002] M. L. Frame and B. B. Mandelbrot. *Fractals, Graphics, and Mathematics Education.* The Mathematical Association of America, Washington, DC, 2002.

[Gardner, 1967] M. Gardner. *An array of problems that can be solved with elementary mathematical techniques.* Sci. Amer., 216:3:124–129, 4:116–123, 6, 1967. (*27*)

[Gardner, 1977] M. Gardner. *Mathematical Magic Show.* Knopf, New York, 1977. (*27*)

[Gleick, 1987] J. Gleick. *Chaos. Making a New Science.* Penguin Books, New York, 1987.

[Hilbert, 1891] D. Hilbert. *Über die stetige Abbildung einer Linie auf ein Flächenstück.* Math. Ann., 38:459–460, 1891. (*30*)

[Hoveijn and Scholtmeijer, 2001] I. Hoveijn and J. Scholtmeijer. *Fractals.* Epsilon Uitgaven, Utrecht, 2001.

[Hutchinson, 1981] J. Hutchinson. *Fractals and self-similarity.* Indiana Univ. Math. J., 30:713–747, 1981.

[Julia, 1918] G. Julia. *Mémoire sur l'itération des fonctions rationnelles.* J. de Math. Pures et Appliquées, 8:47–245, 1918. (*112*)

[Lebesgue, 1905] H. Lebesgue. *Sur les fonctions représentables analytiquement.* Journ. de Math. (6), 1:139–216, 1905. (*39*)

[Lévy, 1938] P. Lévy. *Les courbes planes ou gauches et les surfaces composées de parties semblables au tout.* J. de l'École Polytechnique III, 7–8:227–247, 249–291, 1938. (*43*)

[Mandelbrot, 1977] B. B. Mandelbrot. *Fractals: Form, Chance, and Dimension.* W. H. Freeman and Co., New York, 1977.

[Mandelbrot, 1980] B. B. Mandelbrot. *Fractal aspects of the iteration $z \to \lambda z(1-z)$ for complex λ and z.* Ann. New York Acad. Sci., 357:249–259, 1980. (*123*)

[Mandelbrot, 1982] B. B. Mandelbrot. *The Fractal Geometry of Nature.* W. H. Freeman and Co., New York, 1982. German Edidion: *Die fraktale Geometrie der Natur.* Birkhäuser, Basel, 1991. (*9, 11, 74, 123*)

[Menger, 1926] K. Menger. *Allgemeine Räume und charakteristische Räume, Zweite Mitteilung: Über umfassenste n-dimensionale Mengen.* Proc. Acad. Amsterdam, 29:1125–1128, 1926. (*50*)

[Peano, 1890] G. Peano. *Sur une courbe, qui remplit toute une aire plane.* Math. Ann., 36:157–160, 1890. (*30*)

[Peitgen and Jürgens, 1990] H.-O. Peitgen and H. Jürgens. *Fraktale: Gezähmtes Chaos.* Carl Friedrich von Siemens Stiftung, München, 1990.

[Peitgen and Richter, 1986] H.-O. Peitgen and P. H. Richter. *The Beauty of Fractals.* Springer, Berlin, 1986. (*135, 142, 150*)

[Peitgen, Jürgens, Saupe, 1992] H.-O. Peitgen, H. Jürgens, and D. Saupe. *Fractals for the Classroom. Part 1: Introduction to Fractals and Chaos. Part 2: Complex Systems and Mandelbrot Set.* Springer, Berlin, 1992. German Edition of Part 1: *Bausteine des Chaos. Fraktale.* RoRoRo, Hamburg, 1998. (*85, 86, 90, 93, 95, 96, 98, 135*)

[Schröder, 1991] M. Schröder. *Fractals, Chaos, Power Laws.* W. H. Freeman and Co., New York, 1991.

[Sierpinski, 1915] W. Sierpinski. *Sur une courbe cantorienne dont tout point est un point de ramification.* C.R. Acad. Paris, 160:302, 1915. (*41*)

[Sierpinski, 1916] W. Sierpinski. *Sur une courbe cantorienne qui contient une image biunivoque et continue de toute courbe donnée.* C.R. Acad. Paris, 162:629–632, 1916. (*41*)

[Vicsek, 1989] T. Vicsek. *Fractal Growth Phenomena.* World Scientific, London, 1989.

[von Koch, 1904] H. von Koch. *Sur une courbe continue sans tangente, obtenue par une construction géométrique élémentaire.* Arkiv för Matematik, 1:681–704, 1904. (*6*)

[Zeitler and Pagon, 2000] H. Zeitler and D. Pagon. *Fraktale Geometrie. Eine Einführung.* Vieweg, Braunschweig, 2000.

Literature on general topics

[Cartan, 1966] H. Cartan. *Elementare Theorie der analytischen Funktionen einer oder mehrerer komplexen Veränderlichen.* Hochschultaschenbücher Bibl. Inst., Mannheim, 1966. (*111*)

[Evans and Gariepy, 1992] L. C. Evans and R. F. Gariepy. *Measure Theory and Fine Properties of Functions.* CRC Press, New York, 1992. (*56, 57*)

[Feigenbaum, 1978] M. J. Feigenbaum. *Quantitative universality for a class of nonlinear transformations.* J. Stat. Phys., 19:25–52, 1978. (*146*)

[Kelley, 1955] J. L. Kelley. *General Topology.* D. Van Nostrand Company, New York, 1955. (*3*)

[May, 1976] R. M. May. *Simple mathematical models with very complicated dynamics.* Nature, 261:459–467, 1976. (*146*)

[Montel, 1927] P. Montel. *Leçons sur les familles normales.* Gauthier-Villars, Paris, 1927. (*113*)

[Nevanlinna and Paatero, 1965] R. Nevanlinna and V. Paatero. *Einführung in die Funktionentheorie.* Birkhäuser, Basel, 1965. (*111*)

[Saks and Zygmund, 1971] S. Saks and A. Zygmund. *Analytic Functions.* Elsevier Publ. Comp., Amsterdam, 3rd edition, 1971. (*113*)

[Verhulst, 1845] P. F. Verhulst. *Recherches mathématiques sur la loi d'accroissement de la population.* Nouveaux mémoires de l'Académie Royale des Sciences et Belles-Lettres de Bruxelles, 18:1–38, 1845. (*146*)

[Verhulst, 1847] P. F. Verhulst. *Deuxième mémoire sur la loi d'accroissement de la population.* Mémoires de l'Académie Royale des Sciences, des Lettres et des Beaux-Arts de Belgique, 20:1–32, 1847. (*146*)

List of symbols

The symbols appear for the first time on the indicated page.

$\mathbb{N}$	natural numbers	3
$\mathbb{Z}$	integers	51
$\mathbb{Q}_4$	4-adic rational numbers in $[0, 1]$	47
$\mathbb{R}$	real line	1
$\mathbb{R}^+$	non-negative reals	51
$\mathbb{R}^n$	n-dimensional EUCLIDEAN space	4
$\mathbb{C}$	complex plane	109
$\mathbb{C}^*$	RIEMANN sphere	110
Re	real part	123
Im	imaginary part	142
$\lvert\cdot\rvert$	absolute value, EUCLIDEAN norm	4
$o(g)$	$\lim \frac{\lvert o(g)(x)\rvert}{\lvert g(x)\rvert} = 0$	7
$O(g)$	$\limsup \frac{\lvert O(g)(x)\rvert}{\lvert g(x)\rvert} < \infty$	7
$f \circ g(z)$	$f(g(z))$	121
$f^{(k)}(z)$	$f(f^{(k-1)}(z))$	1
$A_{(k)}$	$f^{(k)}(A)$	1
$\overline{A}$	closure of A	54
∂A	boundary of A	100
$\mathcal{L}, \mathcal{L}^1$	LEBESGUE measure in $\mathbb{R}$	2
$\mathcal{L}^2$	LEBESGUE measure in $\mathbb{R}^2$	45
$\mathcal{L}^3$	LEBESGUE measure in $\mathbb{R}^3$	50
$\dim_S$	self-similarity dimension	8
$\overline{\dim}_B$	upper box counting dimension	51
$\underline{\dim}_B$	lower box counting dimension	51
$\dim_B$	box-counting dimension	51
$\dim_H$	HAUSDORFF dimension	58
$\mathcal{H}^1$	one-dimensional HAUSDORFF measure	37
$\mathcal{H}^s_\delta$	δ-approximation of $\mathcal{H}^s$	56
$\mathcal{H}^s$	s-dimensional HAUSDORFF measure	56
$\alpha(s)$	$\mathcal{H}^s$ of s-dimensional unit ball	55
Γ	gamma function	55
diam	diameter	37
N_δ	minimal cardinality of a δ-covering	50
$\mathcal{K}(X)$	space of compact subsets of X	63
$d(a, b)$	distance between a and b	64
$d(a, B)$	distance of a from B	64
$d(A, B)$	distance of A from B	64
$h(A, B)$	HAUSDORFF distance	64
$B_{\{\varepsilon\}}$	closed ε-hull of B	65

Index

Contents (detailed)

Bei Fragen zur Produktsicherheit wenden Sie sich bitte an:
If you have any questions regarding product safety,
please contact:

Walter de Gruyter GmbH
Genthiner Straße 13
10785 Berlin
productsafety@degruyterbrill.com